I0482695

Disclaimer

Book Title: 3D Imaging Systems for Manufacturing, Construction, and Mobility (NIST TN 1682)

Book Author: Geraldine S. Cheok; Maris Juberts; Marek Franaszek;

Book Abstract: The National Defense Center of Excellence for Industrial Metrology and 3D Imaging (COE-IM3DI) is conducting research to enable the next generation of manufacturing-centric human/instrument/machine interaction. The generalized goal of the COE-IM3DI research is to facilitate the transition of 3D imaging technology from one requiring highly skilled/specialized technicians to a ubiquitous measurement capability available in real-time on the shop floor. This long-term effort requires a detailed examination of the current state of the art in 3D imaging and projected technology development trends over the next ten years. In 2004, NIST produced a seminal report on the then state of the art of the 3D Imaging field with a particular focus on hardware requirements for applications in manufacturing, autonomous vehicle mobility, and construction. This report will extend that initial work (principally with respect to software) and provide an update that will support the needs of the COE-IM3DT research program.

Citation: NIST TN - 1682

Keywords: imaging; hardware survey; focal plane arrays (FPAs); LADAR; laser scanners; LIDAR; software survey

NIST TN 1682

3D Imaging Systems for Manufacturing, Construction, and Mobility

Geraldine Cheok

Maris Juberts

Marek Franaszek

Alan Lytle

National Institute of
Standards and Technology
U.S. Department of Commerce

NIST TN 1682

3D Imaging Systems for Manufacturing, Construction, and Mobility

Geraldine Cheok
Engineering Laboratory

Maris Juberts
Engineering Consultant

Marek Franaszek
Engineering Laboratory

Alan Lytle
NIST Guest Researcher

December 2010

U. S. Department of Commerce
Gary Locke, Secretary

National Institute of Standards and Technology
Patrick D. Gallagher, Director

DISCLAIMER

Certain trade names and company products are mentioned in the text or identified in an illustration in order to adequately specify the experimental procedure and equipment used. In no case does such an identification imply recommendation or endorsement by the National Institute of Standards and Technology, nor does it imply that the products are necessarily the best available for the purpose.

Abstract

The National Defense Center of Excellence for Industrial Metrology and 3D Imaging (COE-IM3DI) is conducting research to enable the next generation of manufacturing-centric human/instrument/machine interactions. The generalized goal of the COE-IM3DI research is to facilitate the transition of 3D imaging technology from one requiring highly skilled/specialized technicians to a ubiquitous measurement capability available in real-time on the shop floor. This long-term effort requires a detailed examination of the current state-of-the-art in 3D imaging and projected technology development trends over the next ten years.

In 2004, NIST produced a report [1] on the then state-of-the-art of the 3D Imaging field with a particular focus on hardware requirements for applications in manufacturing, autonomous vehicle mobility, and construction. This report will extend that initial work (principally with respect to software) and provide an update that will support the needs of the COE-IM3DT research program. The specific 3D imaging areas examined include:

- Current state-of-the-art software and future software trends for 3D image data
- Current state-of-the-art hardware and future hardware trends for active 3D imaging systems
- Assessing operational requirements of 3D imaging for national defense with a focus on manufacturing, construction, and autonomous mobility

Keywords: 3D imaging; hardware survey; focal plane arrays (FPAs); LADAR; laser scanners; LIDAR; software survey.

Table of contents

1 Introduction

The National Defense Center of Excellence for Industrial Metrology and 3D Imaging (COE-IM3DI) is conducting research to enable the next generation of manufacturing-centric human/instrument/machine interactions. The generalized goal of the COE-IM3DI research is to facilitate the transition of 3D imaging technology from one requiring highly skilled/specialized technicians to a ubiquitous measurement capability available in real-time on the shop floor. This long-term effort requires a detailed examination of the current state-of-the-art in 3D imaging and projected technology development trends over the next ten years. This proposed research by the National Institute of Standards and Technology (NIST) would directly address this need.

The Construction Metrology and Automation Group (CMAG) at NIST collaborated with Youngstown State University in support of the COE-IM3DI to produce a detailed review of the 3D imaging research and development field, provide an in-depth understanding of the current state-of-the-art in active-sensor 3D imaging, and project development trends over the next decade. In 2004, NIST produced a then state-of-the-art report [1] of the 3D Imaging field with a particular focus on hardware requirements for applications in manufacturing, autonomous vehicle mobility, and construction. This report will extend that initial work (principally with respect to software) and provide an update that will support the needs of the COE-IM3DT research program. The specific 3D imaging areas examined include:

- Current state-of-the-art software and future software trends for 3D image data
- Current state-of-the-art hardware and future hardware trends for active 3D imaging systems
- Assessing operational requirements of 3D imaging for national defense with a focus on manufacturing, construction, and autonomous mobility.

3D imaging (3DI) systems are used to collect large amounts of 3D data of an object or scene in a short period of time. A definition of a 3D imaging system as taken from ASTM E2544 [2] is:

A 3D imaging system is a non-contact measurement instrument used to produce a 3D representation (for example, a point cloud) of an object or a site.

DISCUSSION—
(1) Some examples of a 3D imaging system are laser scanners (also known as LADARs [Laser Detection and Ranging] or LIDARs or laser radars), optical range cameras (also known as flash LIDARs or 3D range cameras), triangulation-based systems such as those using pattern projectors or lasers, and other systems based on interferometry.
(2) In general, the information gathered by a 3D imaging system is a collection of n-tuples, where each n-tuple can include but is not limited to spherical or Cartesian coordinates, return signal strength, color, time stamp, identifier, polarization, and multiple range returns.
(3) 3D imaging systems are used to measure from relatively small scale objects (for example, coin, statue, manufactured part, human body) to larger scale objects or sites (for example, terrain features, buildings, bridges, dams, towns, archeological sites).

The work in this report was conducted under the CRADA (Cooperative Research & Development Agreement) ID number CN-09-5118 between NIST and Youngstown State University and a contract between Mr. Maris Juberts, an engineering consultant and a former NIST employee, and Youngstown State University.

2 Current Applications for 3D Imaging Systems

2.1 Manufacturing

Applications in manufacturing for active 3D imaging systems include: tooling and product/part inspection, capturing physical geometry of parts, reverse engineering of parts and assemblies, CAD part modeling, capturing production line equipment and manufacturing sites in 3D for documentation and production line planning, large volume metrology, navigation and collision avoidance for Automated Guided Vehicles (AGV), safety and security systems for automated equipment in assembly line operations. Table 1 contains a list of the main features and salient characteristics of active 3D imaging methods used in manufacturing applications.

Table 1. Typical Applications in Manufacturing

Type of 3D imaging methods	Main features & salient characteristics	Typical range for uncertainty values	Typical applications
Triangulation (active)	1D, 2D, & 3D illumination Depth of field (DOF) < 4 m[*] High measurement rate (10 to 1000) kHz[*] Most require scanning Some shadowing Low cost	0.02 mm to 2.00 mm[*]	Reverse engineering, quality assurance, CAD part modeling, part/assembly inspection
Pulsed Time-of-Flight (TOF) (scanning)	light beam illumination DOF > 10 m[*] Requires scanning Good measurement rate (1 to 250) kHz[*] Good in bright ambient (outdoors) Multiple pulse return detection-possible	5 mm to 50 mm*	Machine, production line, building, infrastructure modeling and documentation
AM (Amplitude Modulated) TOF-(scanning)	light beam illumination DOF: (1 to 100) m[*] Requires scanning High measurement rate (50 to 1000) kHz[*] May be affected by bright ambient light Primarily used indoors	0.05 mm to 5.00 mm*	Tooling, reverse engineering, inspection, metrology of larger parts and assemblies. CAD part modeling. Machine, industrial site & infrastructure modeling and documentation.
FM (Frequency Modulated) TOF-(scanning)	light beam illumination DOF < 10 m[*] Slow scanning rate (0.01 to 2.0) kHz[*] Highest detection sensitivity Lowest power requirements High dynamic range Most expensive	0.01 mm to 0.25 mm*	large scale dimensional metrology

Type of 3D imaging methods	Main features & salient characteristics	Typical range for uncertainty values	Typical applications
AM TOF Focal Plane Array	Entire FOV Illumination Non-scanning Narrow FOV 40° x 40° typical Largest array size 200 x 200 Real-time operation up to 100 frames per second (fps) Typical DOF: (0.5 to 30) m Limited operation outdoors. Low cost	±1 cm	Navigation & safety systems for AGVs & other mobility platforms. Real-time perception for industrial robots & assembly line equipment and safety systems.
Confocal Optical Coherence Tomography & White Light Interferometry	Flying spot scanning in two dimensions – relatively slow Broadband light source – low coherence. Short standoff distances & limited to a few millimeters of depth imaging	< ±10 µm	Subsurface 3D imaging of translucent or opaque materials. Non-destructive material tests. Reverse engineering of small parts.
Parallel Optical Coherence Tomography (pOCT)	Uses CCD camera or CMOS smart pixel arrays. Fast measurement rate: up to 10 3D fps. Broadband light source – low coherence. Short standoff distances & limited to a few millimeters of depth imaging	< ±10 µm	Subsurface 3D imaging of translucent or opaque materials. Non-destructive testing of parts and materials. Reverse engineering of small parts – such as MEMS devices.

* Beraldin, J.A., "Basic Theory on Surface Measurement Uncertainty of 3D Imaging Systems," Proceedings of SPIE-IS&T Electronic Imaging, Vol. 7239, 2008.

2.2 Construction

The majority of the 3DI systems used in construction are time-of-flight instruments. Common applications include surveying, volume determination, clash detection, creating as-built models, dimensional checking, tolerance checking, and topographic mapping. The use of 3DI systems has the added benefit of allowing the work to be performed at a safe distance, e.g., bridge or road surveying.

Typical characteristics of 3DI systems used for construction applications are:

- horizontal fields-of-view (FOVs) of 360°
- vertical FOVs of 80° or better
- measurement uncertainties on the order of millimeters
- maximum ranges from about 50 m to over a kilometer.

3DI systems have been used by various state Departments of Transportation (DOTs) for land and highway surveys. The 3DI systems used for these types of applications have maximum ranges of

100 m to 200 m. For volume determination (e.g., mining, excavations) and topographic mapping, systems with longer maximum ranges are more efficient. 3DI systems used for clash detection and for the creation of as-built models typically have maximum ranges of 50 m to 200 m. Systems used for tolerance checking have ranges of 50 m or greater and measurement uncertainties on the order of millimeters or better.

2.3 Autonomous Mobility

Applications in mobility systems for active 3D imaging systems include: Automated Guided Vehicle and other industrial vehicle navigation and collision avoidance; military autonomous Unmanned Ground Vehicle (UGV) navigation, terrain mapping and collision avoidance; automotive collision avoidance, mapping and surveying. Table 2 contains a list of the main features and salient characteristics of active 3D imaging methods used in mobility systems applications.

Table 2. Typical Applications in Mobility Systems.

Type of 3D imaging methods	Main features & salient characteristics	Typical range for uncertainty values	Typical applications
Pulsed TOF (scanning)	Light beam illumination Multiple scanning planes & laser/ detector pairs Wide FOV available DOF >100 m Measurement rate > 1 million pixels per second Good in bright ambient light (outdoors) Capable of multiple pulse return detection	5 mm to 50 mm	Military UGV autonomous navigation, terrain mapping and collision avoidance. Automotive collision avoidance Mapping and surveying.
Pulsed TOF (non-scanning) Focal Plane Array	3D illumination-high power Non-scanning Variable narrow FOV DOF > 10 m up to kilometers Up to 256 x 256 array size Measurement rate typically > 30 Hz Good in bright ambient light (outdoors) Capable of multiple pulse return detection	2 cm to 5 cm	Robotic vehicle navigation, mapping, security and targeting.
AM TOF (non-scanning) Focal Plane Array	Entire FOV Illumination Non-scanning Narrow FOV 40°x40° typical Largest array size 200 x 200 Real-time operation up to 100 fps Typical DOF: (0.5 to 30) m Limited operation outdoors Low cost	± 1 cm	Navigation & safety systems for AGVs & other mobility platforms. Real-time perception for industrial robots & assembly line equipment and safety systems. Automotive safety systems.

3 Current State-of-the-Art (2010)

3.1 Introduction

A 3D imaging project involves the following basic steps:

1. Selecting a location that minimizes occlusions.
2. Setting the instrument on a stand or tripod. Depending on the instrument, the instrument can be attached to a tribrach and centered over a point below the tripod.
3. Collecting data from several locations – see Figure 1. This is necessary since a 3DI system is a line-of-sight instrument and thus, data from several different locations are needed to obtain a full 3D model. The data is stored in the laptop/PC that controls the instrument or on memory cards in the instrument.
4. Exporting the data.
5. Registering the data. Registration is defined as the process of determining and applying to two or more datasets the transformations that locate each dataset in a common coordinate system so that the datasets are aligned relative to each other [2]. Sometimes registration to a global reference frame (e.g., state plane) is required in a project.
6. Processing data. *"Final data processing can represent a significant effort on some projects--a structure that is scanned in 4 hours may take as much as 40 hours or longer to process the data to a CADD compatible format. The software used for processing scanned datasets is a critical component in the overall efficiency and economy of the process. [3]"*

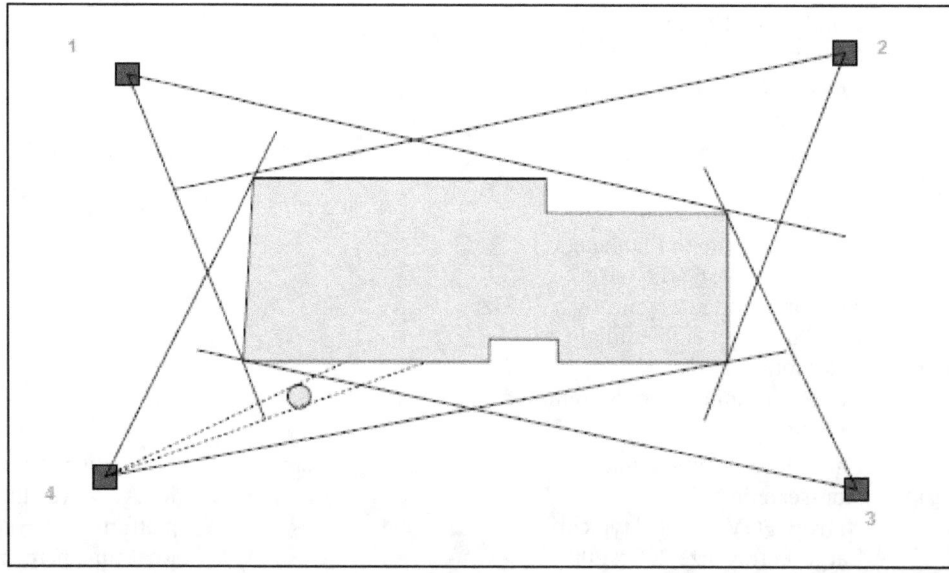

Figure 1. Multiple scan locations (indicated by the small square boxes) are needed to fully model a building and cover obscured areas. In practice, additional scan locations may be needed due to occlusions from vegetation or other objects between the building and the scanning instrument. Figure taken from [3].

A general workflow for a 3D imaging project is shown in Figure 2. The hardware primarily affects the planning and data collection stages and the software primarily affects the data processing, data analysis, and modeling stages. The hardware and software are described in more detail in the remaining sections of this chapter.

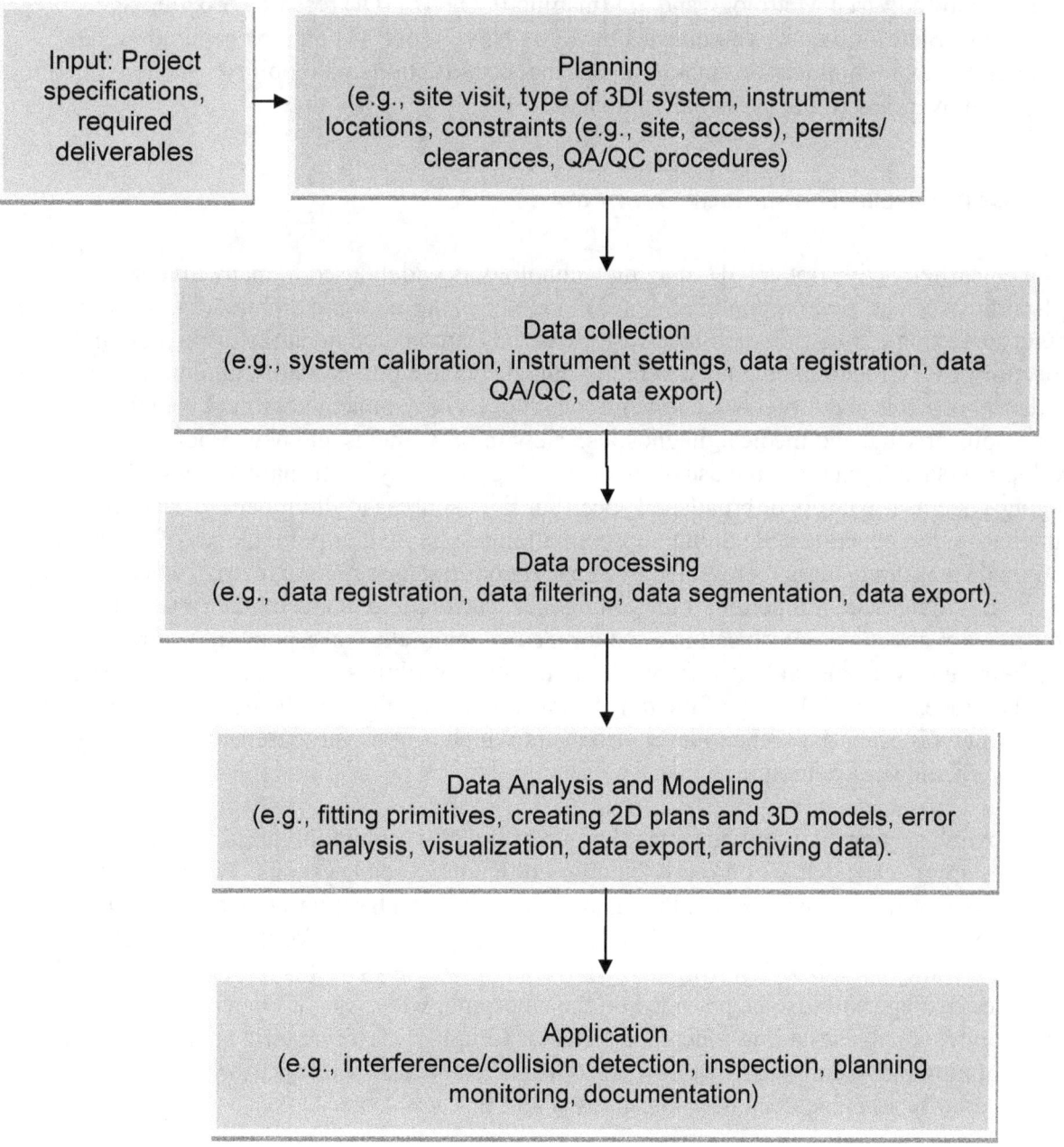

Figure 2. Simplified workflow for a 3D imaging project.

3.2 Active 3D Imaging Hardware

This portion of Chapter 3 covers the hardware portion of the current state-of-the-art and ongoing research in active 3D imaging technology as it currently exists for a variety of industrial, automotive and military applications. It supports the needs of the National Defense Center of Excellence in Industrial Metrology and 3D Imaging (COE-IM3DI) research program at Youngstown State University and updates the 2004 NIST report [1] on next generation 3D imaging systems for applications in manufacturing, construction and mobility. Section 3.2.1 provides an overview of the hardware approaches.

3.2.1 Commercial Technology Overview

As stated earlier, active optical 3D imaging technology is widely used in many diverse applications such as: reverse engineering (3D modeling), heritage and architectural preservation, ground surveying (terrestrial, mobile terrestrial, aerial), automated manufacturing (assembly, inspection), AGV (automated guided vehicle) guidance and safety systems, automotive collision avoidance, obstacle and target detection and tracking, UGV (unmanned ground vehicle) autonomous driving, and medical diagnostics. New improvements and new product developments could increase the use of this technology in many other market areas. Active 3D imaging systems use lasers or broadband spectrum light sources to illuminate surfaces in a scene of interest and to generate a 3D digital representation (range map or point cloud) of the surface. Each range measurement is represented in a spatial coordinate system (a matrix) where the row and column indices are a function of the orthogonal scan angle or some regular interpolated grid in the x and y directions. Each cell in the matrix can contain the corresponding depth measurement (z values), calibrated (x, y, z), or any other attributes such as color or uncertainty [4]. The processing and display of the range data can directly provide the geometry of the surface of an object or the scene in an environment which is relatively insensitive to background illumination and surface texture.

The following sections will provide a breakdown of the most commonly used measurement principles/approaches that are currently being used in commercial systems. Because of the limited scope of the hardware study, the main focus will be on laser based optical triangulation approaches and optical time delay (TOF measurement) approaches. For the same reasons, scanning techniques and devices will not be covered. Overviews of this topic can be found in [1, 5]. Some coverage will also be provided on the emerging, wider use of Optical Coherence Tomography, which uses a low coherence (broadband light) interferometric approach for very high resolution range imaging. A survey of some currently available hardware products and research prototypes is included in Appendices A and B.

3.2.1.1 Breakdown of Optical 3D Imaging Hardware by Range Measurement Techniques

The most common optical active 3D range imagers utilize the basic measurement principles of (1) triangulation, (2) time-of-flight, and (3) coherent time-of-flight or interferometry. Figure 3 shows the family of optical 3D range/shape measurement techniques. These techniques mostly use light wave illumination sources in the wavelength range of 400 nm to 1600 nm (visible and near infrared (NIR) spectrum). This section will summarize the basic measurement principles of each approach. Microwave and ultrasonic wave techniques will not be included in the description.

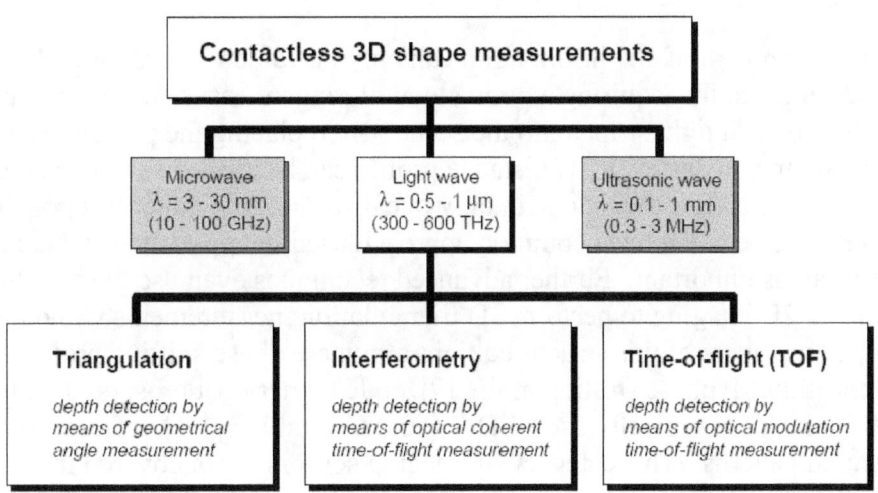

Figure 3. Family tree of non-contact 3D shape measurement techniques [6].

3.2.1.1.1 Triangulation

Laser-based, triangulation scanning represents the most commonly used approach in industrial applications. Figure 4 illustrates the principle of active triangulation range measurement.

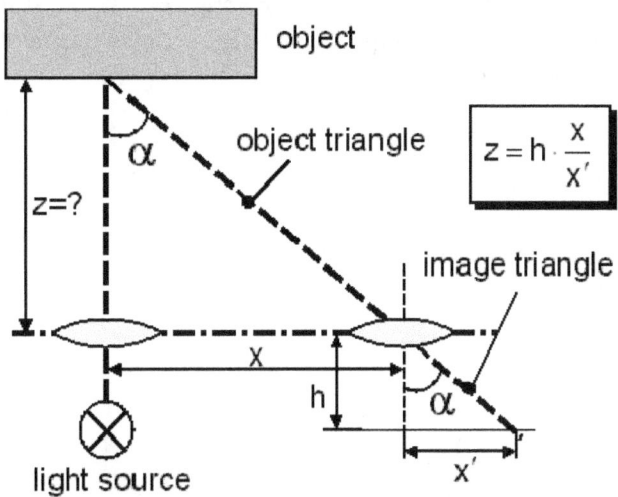

Figure 4. Principle of active triangulation range measurement [7].

9

Laser-based, triangulation 3DI systems project a laser spot or stripe onto the object being measured and then use a linear or 2D array detector to locate the illuminated area on the object being inspected or measured. The light source, the illuminated area on the object, and the detector form a triangle as illustrated in Figure 4. Lange provides a basic overview of the laser-based triangulation measurement technique in his 2000 doctoral dissertation [7]. Much of the following text comes from that dissertation.

With a laser spot projection, 3D information can only be gained by scanning the laser spot over the scene and sequentially acquiring a point cloud of range measurements. A much faster approach is to project a light stripe onto the scene. By replacing the position sensitive linear detector with an imaging array, a 2D distance profile can be measured with a single measurement. The scan has to be done in only one direction to obtain a 3D point cloud of the scene. The laser stripe is the most common approach used in applications where fast measurement rate is important. Further advanced techniques even use 2D structured light illumination and 2D imaging to perform 3D triangulation measurements without scanning. The most important members of this structured light group are phase shifting projected fringe, greycode approach [8], phase shifting moiré [9], coded binary patterns, random texture [10] or color-coded light. They typically use LCD (Liquid Crystal Display) projectors for the projection of the structured patterns. This category of imaging sensors is not covered in this report.

3.2.1.1.2 Time-of-Flight Techniques

Light waves travel at a velocity (c) of approximately $3*10^8$ m/s through the earth's atmosphere. There is a class of measurement methods which take advantage of this knowledge by measuring the time delay created by light traveling from an illumination source to a reflective target and back to a detector located in close proximity to the position of the illumination source. The light beam travels a distance which is twice that of the distance to the reflective target. Figure 5 illustrates the most basic principle of a time-of- flight (TOF) ranging system (a pure "pulsed" time-of-flight range measurement principle). A very short laser pulse (0.5 ns to 5 ns) is used to start a timing circuit. Once the detector picks up the reflected signal, it immediately stops the timer. An extremely accurate clock is required to achieve measurements with small errors. The distance, d, to an object is equal to (c * TOF delay divided by 2).

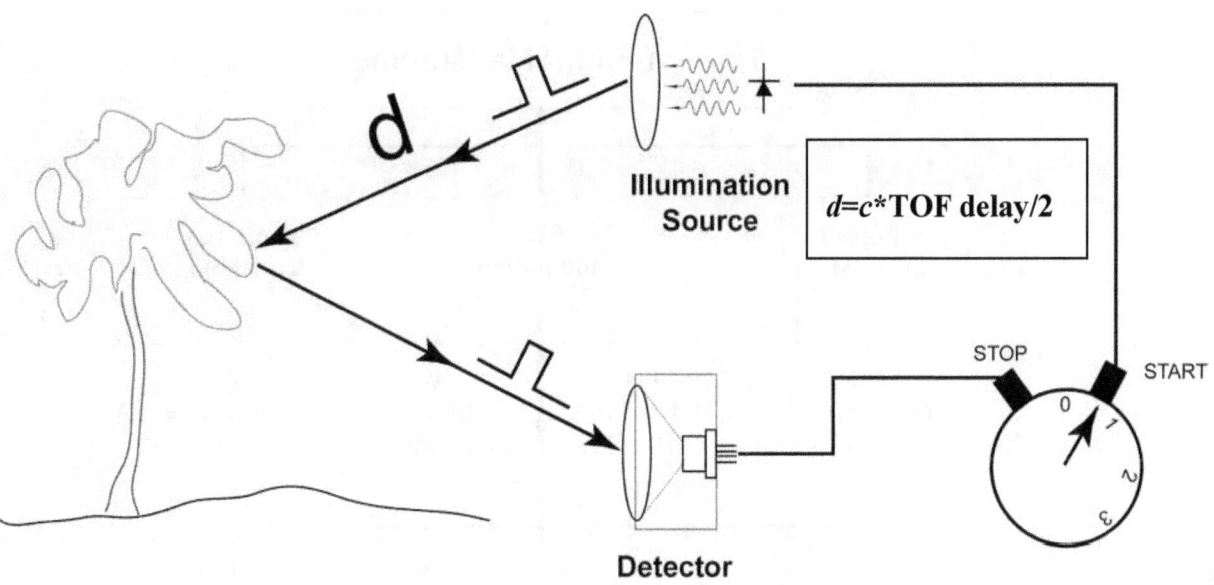

Figure 5. Fundamental principle of a "pulsed" time-of-flight range measurement technique (adapted from Lange [7]).

Time-of-flight range measurement systems use either pulsed modulation or continuous wave (CW) modulation of the illumination source. Figure 6 shows a "family tree" of the different types of modulation signals used in time-of-flight range measurement systems. The three main categories are: pulsed modulation; CW modulation and pseudo-noise modulation. Pseudo-noise modulation will not be covered here.

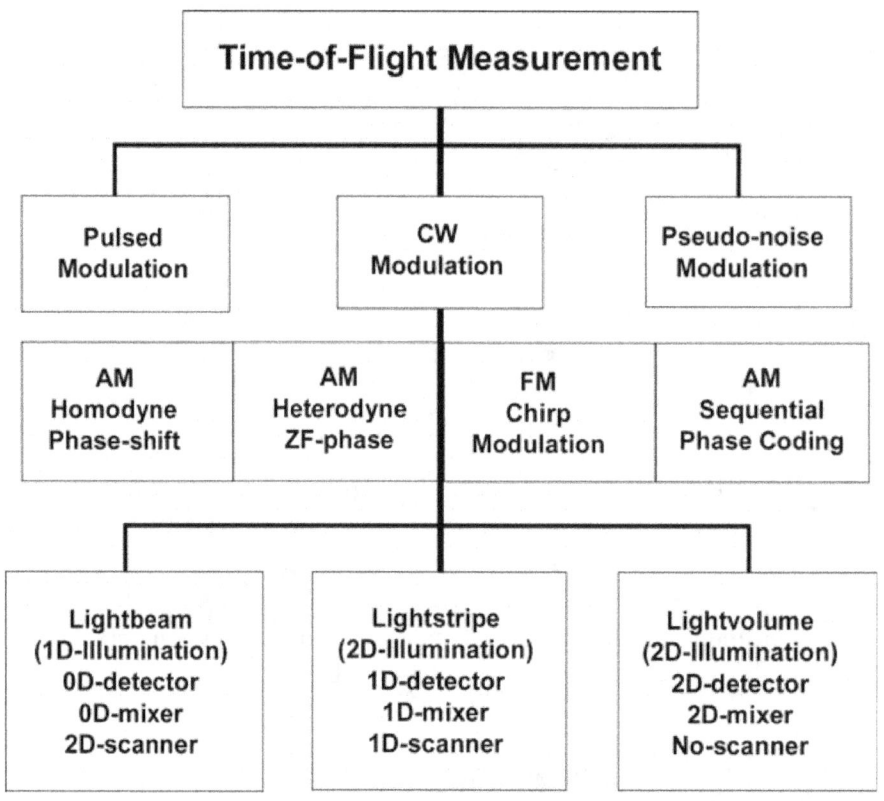

Figure 6. "Family tree" of the different types of modulation signals used in Time-of-Flight range measurement systems. (adapted from Lange [7]).

Continuous wave modulation techniques can be further broken down into the following sub-categories: Homodyne Amplitude Modulation (AM) (single modulation frequency) phase shift techniques; Heterodyne AM (multiple modulation frequencies); Frequency modulation chirp (linear frequency sweep of the coherent wavelength source) modulation; and AM sequential phase coding [11]. The technical details for these techniques will be described in the following section.

3.2.1.1.3 Interferometry time-of-flight range measurement techniques

Single or multiple wavelength interferometry and Optical Coherence Tomography (OCT) can be classified separately as a third active range measurement technique. The most basic interferometer setup (the Michelson interferometer) is illustrated in Figure 7.

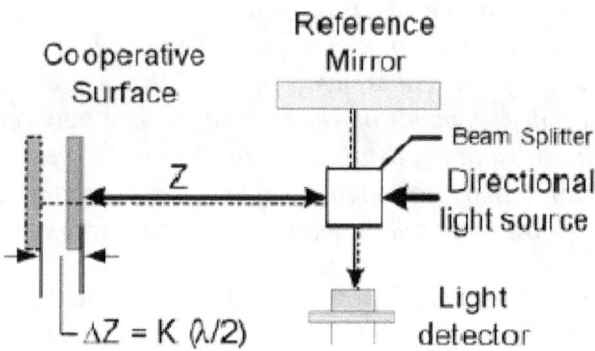

Figure 7. Working principle of Michelson interferometer (adapted from Beraldin [4]).

The following basic overview of the interferometric technique comes from the Lange dissertation [7]. In the figure, a monochromatic and coherent laser beam is split into two rays by a beam splitter. One ray is projected to a reference mirror (constant reference path) while the other is projected on the object/target (variable measurement path). Both rays are reflected back to the beam splitter, which projects the rays onto the detector electronics where the interference is measured and counted. A movement of the object away or towards the sensor results in a constructive interference peak being observed each time the object moves by a multiple of the laser's half wavelength. By counting the min-max transitions in the interference pattern, the relative distance of object motion can be determined at an uncertainty on the order of the laser light's wavelength (λ) or better. The technique can be interpreted as a time-of-flight principle, since the runtime (time delay) difference between the reference and measurement path is evaluated. Lange also provides the following overview of other interferometric approaches and associated references in his dissertation [7]:

> "*Several other interferometer setups can be found, for example in [12] or [13]. The principal drawback of classical interferometers is that absolute distance measurements are not possible and the unambiguous distance range is as low as half the wavelength. Enhanced approaches overcome this restriction. One nice example is Multiple-wavelength interferometry, where two very closely spaced wavelengths are used at the same time. [This allows] beat frequencies down to GHz or even MHz range [to be] synthetically generated, enabling absolute measurements over several tens of centimeters at $\lambda/100$ resolution [6, 14]. Especially important for high sensitivity 3D deformation measurements is the electronic speckle pattern interferometry (ESPI), where a reference wave front interferes with the speckle pattern reflected by the investigated object. With a conventional CCD camera the speckle interferogram, carrying information of the object deformation, can then be acquired. Like conventional interferometers, ESPI can also be improved in sensitivity and measurement range by the multiple-wavelength technique [14].*
>
> *Another way to enlarge the measurement range is to use light sources of low coherence length. Such interferometers (white-light interferometry or low-coherence interferometry*

[15]*) make use of the fact that only coherent light shows interference effects. If the optical path difference between the measurement and reference paths is higher than the coherence length of the light, no interference effects appear. For a path difference of the order of magnitude of the coherence length, however, interference takes place. The strength of interference, depends on the path difference between the reference and object beams, and thus absolute distances can be measured. Interferometry finds its applications predominantly for highly accurate measurements ($\lambda/100$ to $\lambda/1000$) over small distances ranging from micrometers to several centimeters."*

Because of the limited scope of the hardware study, further detailed technical description of the traditional/classical interferometry approaches is not be provided here. An overview on these techniques can be found in the 1999 Handbook of Computer Vision and Applications [16]. On the other hand, since low-coherence or Optical Coherence Tomography (OCT) has generated a lot of interest in the medical diagnostics community (especially in ophthalmology applications) and now in manufacturing applications, further information and references will be provided. The following material is a summary of information on OCT available at: http://en.wikipedia.org/wiki/Optical_coherence_tomography.

This coherent approach uses either a white (broadband) light source (such as super bright LEDs) or femtosecond laser pulses and uses short/low coherence interferometry to generate interference distances of only a few micrometers. The light source is used to penetrate through opaque or translucent material or tissue samples to generate reflectance range data to resolutions better that 10 μm. Light interaction is broken into two paths – a sample path and a reference path. The reflected light from the sample and the reference signal are optically mixed to produce an interference pattern. The interferometer strips off any scattered light outside the short coherence interference length. Because of the exceptionally high range resolution, this approach has gained acceptance in the manufacturing community – especially where it is necessary to obtain precise 3D imaging inside translucent or opaque materials. To generate 2D data sets (cross section image of the object) a scanning of the object can be accomplished by using a translation stage to move the object. This is also called single (confocal) OCT. To generate a 3D data set (volumetric image) an area scan must be done. This can be accomplished by using a CCD camera as the detector (does not need an electro-mechanical scan). By stepping the reference mirror and taking successive scan images, a 3D volumetric image can be constructed. The only limitation of this approach is the short standoff distances and imaging of less than 3 mm below the surface of opaque materials. Additional information on OCT technology can be found in [17, 18].

3.2.1.2 <u>Technical review of most common range measurement approaches used commercially</u>

3.2.1.2.1 Laser scanning techniques for active triangulation range imaging

The National Research Council (NRC) of Canada was one of the first organizations to invent and develop the triangulation based laser scanning approach [19]. Since then they have published many papers on the topic, which are available as NRC documents. A summary of the basic principles of triangulation systems is in this report. Technical details can be found in the following NRC and other publications [4 , 20-24]. Beraldin , et al. [23] provides an explanation

14

of the basic principles in the triangulation range measurement concept. The following is a summary of the principles, applications and limitations of the approach presented in [23].

The triangulation approach for 3D imaging is primarily used in applications where the range to the object being scanned is less than 10 m. There are three categories: (1) short range imagers for standoff distances of less than 0.5 m, (2) medium sensors for distances of 0.5 m to 2 m, and (3) larger sensors for distances between 2 m to 10 m.

In basic triangulation, the angles of the similar triangles which are formed by the object triangle and image triangle are measured indirectly by using the ratio of two distances (based on law of cosines). This ratio is shown in Figure 4 to be

$$\frac{z}{h} = \frac{x}{x'}$$

since the angle α is the same in both triangles,

$$z = h \times \frac{x}{x'} \qquad\qquad \text{Eq. 1}$$

The basic measurement principles of active triangulation imaging is shown in Figure 8 and Figure 9 where Figure 8 shows a single laser spot range measurement and Figure 9 depicts the approach used for object scanning.

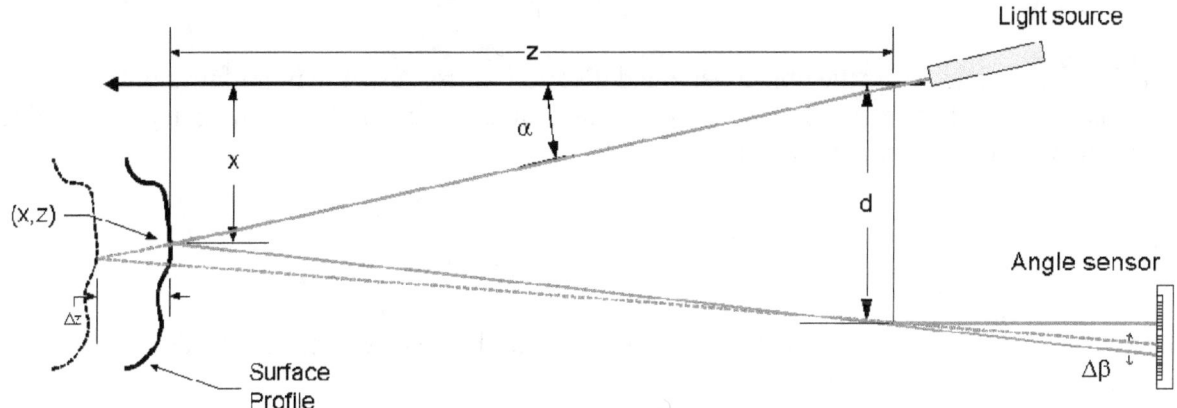

Figure 8. Active triangulation based single spot laser range measurement. [23].

15

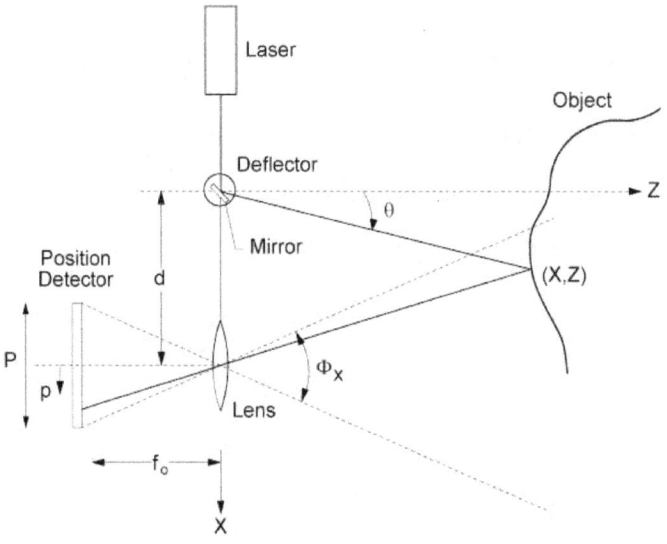

Figure 9. Active triangulation based scanning type laser ranging. From NRC publication [23].

For a single spot laser range measurement , as shown in Figure 8, the shape of the similar triangles are determined from angles α and $\Delta\beta$ relative to the baseline distance d. In Eq. 1, substituting d for x, f_0 for h, and p for the measured position x' on the detector results in the Z equal to

$$ Z = f_0 \times \frac{d}{p}. $$

If the laser is scanned over the object using a deflection mirror, as shown in Figure 9, the coordinates of the z and x measurement can be calculated using simple trigonometry. The value of z for this configuration is now

$$ Z = f_0 \times \frac{d}{(p + f_0 \tan\theta)} $$

The additional term $f_0 \tan \theta$ in the denominator is due to the displacement on the detector from the scanning angle θ in the object triangle.

The NRC paper continues by describing some of the limitations of the scanning triangulation method. The standard deviation of error in the z value, σ_z, is given by

$$ \sigma_z = \frac{z^2}{(f_0 d)} \sigma_p $$

where σ_p is the standard deviation of the error in measuring p. It appears that the uncertainty in z can be lowered by increasing the detector focal length f_0 and the baseline distance d.

16

Unfortunately, this is misleading because shadow effects increase and overall stability of the sensor structure decreases with increasing values of d. Also, since the field of view φ_x of a conventional triangulation sensor is given by

$$\varphi_x = 2\tan^{-1}\left(\frac{P}{2f_0}\right)$$

where P is the length of the position detector, a compromise has to be made when selecting the parameters for a desired application and performance. The paper points out that triangulation sensing performance can be improved by using a scanning scheme that was developed by Rioux [20] that allows a large FOV to be achieved even with small triangulation angles. This scheme is illustrated in Figure 10. With this approach, the instantaneous FOV of the position detector follows the deflection angle of the laser beam. A sensor using this scheme can achieve a wide FOV with small optics and a small baseline.

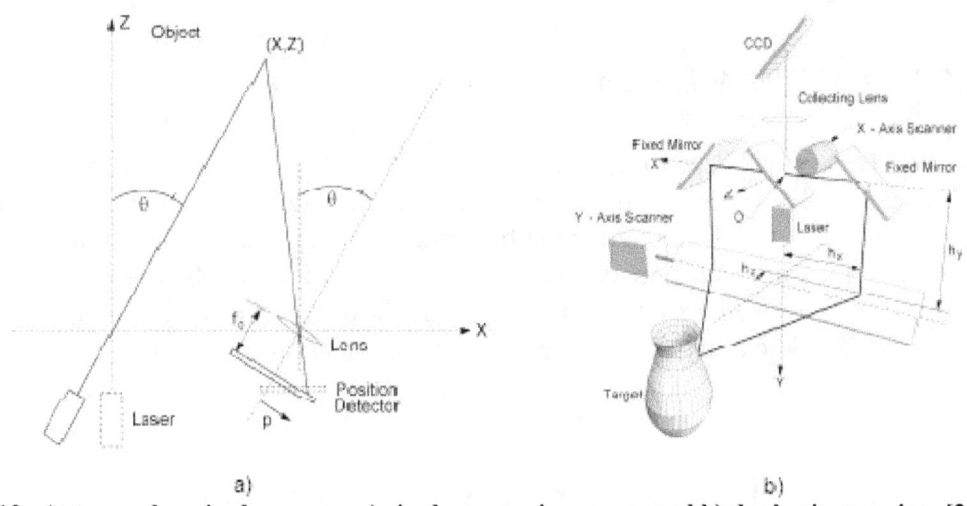

Figure 10. Auto synchronized scanner: a) single scan axis scanner and b) dual axis scanning. [23].

When using laser triangulation scanners, full 3D image data sets can be achieved by moving the scanner or by using galvanometers to drive the laser beam displacement mirrors along the x- and y-axes in order to produce a raster range image. For metrology applications, laser line scanners are typically mounted on Coordinate Measuring Machines (CMMs); however, if lesser performance is acceptable, they can be mounted on a rotation stage or a translation stage to achieve full 3D imaging.

In another more recent NRC publication [4], Beraldin describes the intrinsic uncertainty of active optical 3D imaging measurement techniques. He states that range measurement uncertainties for each class of active 3D imagers can be represented by a single equation that depends on the signal-to-noise (SNR) ratio. If SNR is assumed to be greater than 10, then the range uncertainty, δ_r, can be shown to be

$$\delta_r \approx \frac{K}{\sqrt{SNR}}$$

K is a constant which depends on the range measurement methodology used. For laser based triangulation scanners

$$K \approx \frac{Z^2}{(f\,d)} \frac{1}{BW}$$

where Z is the distance to the surface being scanned, f is the focal length of the detector collector lens, d is the sensor baseline distance, and BW is root-mean-square signal bandwidth. Although speckle noise does affect active triangulation based measurement performance when SNR is very large, in most cases, performance (range of operation) is primarily limited by the geometry of the sensor. Typical values of range uncertainty for active triangulation-based 3D imaging sensors is between 0.02 mm to 2 mm. In addition, the paper states that the depth-of-field (DOF) and measurement rates for these sensors are typically less than 4 m and between (10 to 1000) kHz, respectively. More in depth description of other contributions to the measurement uncertainty can be found in publications [25-27].

3.2.1.2.2 Time of Flight Measurement Techniques

The 2004 NIST report [1] describes in detail the fundamental time-of-flight measurement techniques used in 3DI systems that were available at that time. This report will not attempt to duplicate this effort but will summarize the fundamental science of each approach, list the advantages and limitations, and cover some of the most common applications. Expanded discussion will be provided on advances being made in focal plane array technology and particularly for real-time applications where either the sensor or the scene is in motion.

(1) Pulsed Modulation (Pure pulse TOF)

A typical setup for direct TOF laser-based 3DI system is illustrated in Figure 11.

18

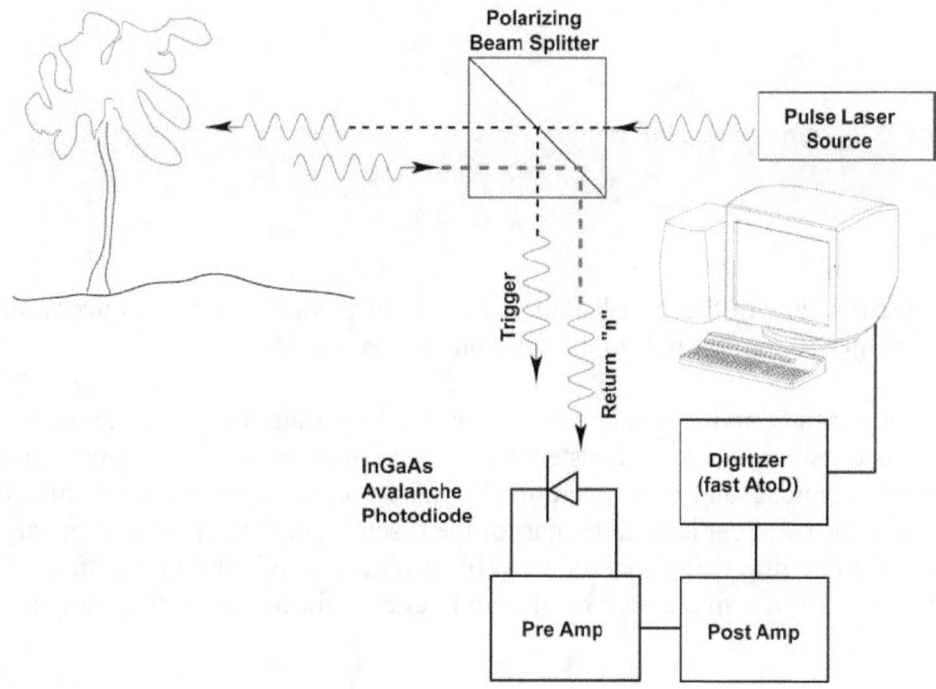

Figure 11. Typical setup for a direct-time-of-flight laser-based 3D imaging system.

One very obvious improvement over the triangulation scheme is that the laser illumination source and the detector operate synchronously, producing full range data sets without any shadowing. Essentially, the illumination beam and detector beam can be made collinear. By keeping the laser pulses very short (a few nanoseconds or even a few hundred picoseconds), a large amount of light energy can be transmitted, thus reducing the effects of background illumination and producing a high signal to noise ratio. The large signal to background illumination ratio reduces the need for the detector to have a high sensitivity and high signal to noise performance. This high signal to background illumination ratio allows for long distance measurements even under bright ambient conditions. Also, the low mean value of the optical power that is used, is extremely important in preventing eye damage, especially if the laser wavelength is in the IR (infrared) range [7].

One drawback of the pulsed TOF measurement approach is the need for the receiver electronics to have a large bandwidth. Currently, the maximum attainable bandwidth that is commercially attainable is limited to about 15 GHz[1]. This means that the achievable range resolution (ΔR) is

$$\Delta R \approx \frac{c}{2}\frac{1}{B}$$

where B is the bandwidth of the receiver electronics. ΔR is typically no better than 1 cm.

Range uncertainty for the various range measurement methods can be expressed as:

---

[1] From a Bridger Photonics presentation.

$$\delta_r \approx \frac{K}{\sqrt{SNR}}$$

The value of K for pure pulse TOF ranging is:

$$K = \frac{c}{2} T_r$$

where T_r is the rise time of the laser leading edge. Typical values of range uncertainty are between 5 mm and 50 mm for pulsed modulation approaches [4].

If low range uncertainty and high range resolution are important, the pulse duration should be kept as short as possible. Some 3DI systems have been operated with laser pulses as short as 250 ps [28-30]. Femtosecond pulse operation has also been demonstrated, but this topic will be described under the coherent laser radar part of the discussion. "Super resolution" algorithms are another way of improving range uncertainty. This is a process of combining information from multiple images to form a single high resolution image. Information on these algorithms can be found in [31] and [32].

There are other limitations that factor into performance of pure TOF 3DI systems:

(a) Laser sources used in generating short duration high optical peak power pulses generally have low repetition rates [typically (1000 to 250 000) pulses/s)]. This tends to limit the acquisition rates of single emitter/single detector scanning devices. The highest repetition rates have been achieved using microchip laser sources in combination with single photon Geiger-mode APDs (avalanche photodiodes) [28-30]. This limitation can be overcome by using large area flash focal plane array (FPA) detectors or by using multiple laser/detector pairs on the scanning mechanism. These will be covered in Section 3.2.2.4 under the topic of real-time active 3DI systems.

(b) Because of imperfect optics and atmospheric dispersion, the illumination beam expands with range. Figure 12 shows the divergence characteristics for an uncompensated microchip laser beam and another for a diffraction-limited optics laser beam. If the unambiguous range of the distance measurement is very large (> 50 m), the reflected return signal will contain multiple target reflections that are within the cone of the dispersion beam. A possible time domain range profile is shown in Figure 13. Most commercial grade pulse imagers average the values of the reflected returns and present only a single measured value. This results in phantom points being generated along vertical and horizontal edges of objects [1]. Some producers, however, offer multiple return features such as: reporting first and last return, more than two returns, or even multiple time slicing and range measurements following an initial detected return [33, 34].

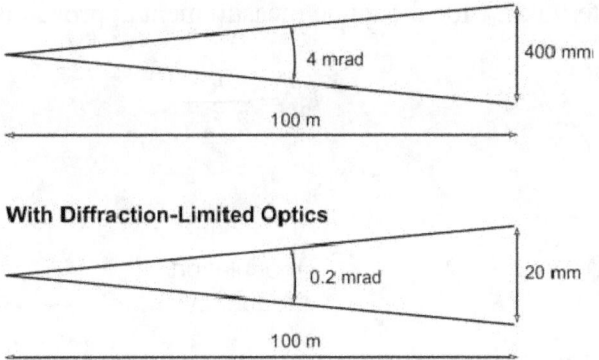

Figure 12. Laser beam divergence and beam diameter at 100 m for an uncompensated MicroChip Laser Beam and for a Diffraction-Limited Optics beam.

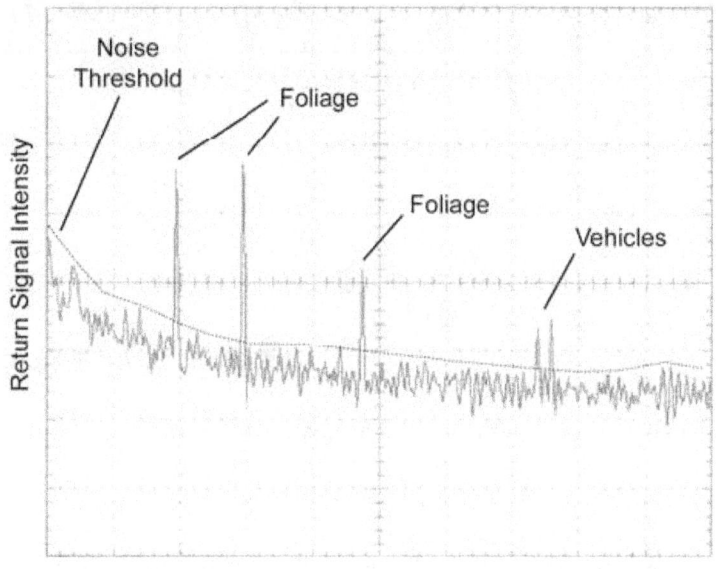

Figure 13. Full time-domain profile return response for a 1ns pulse imager (courtesy Night Vision Lab).

(2) Amplitude Modulated Continuous Wave (AM-CW) (phase–based measurements)

Another way of measuring TOF is by modulating the laser beam with a continuous wave signal and then measuring the phase difference between the emitted signal and the reflected/return signal. This measurement approach is represented in Figure 14. The distance d to the object is given by:

$$d = \frac{\Delta\emptyset \, c}{4 \, \pi f_{AM}}$$

where $\Delta\phi$ is the phase difference and f_{AM} is the modulation frequency.

21

The unambiguous interval, L_u, for the range measurement approach is given by:

$$L_u = \frac{c}{2\,f_{AM}}$$

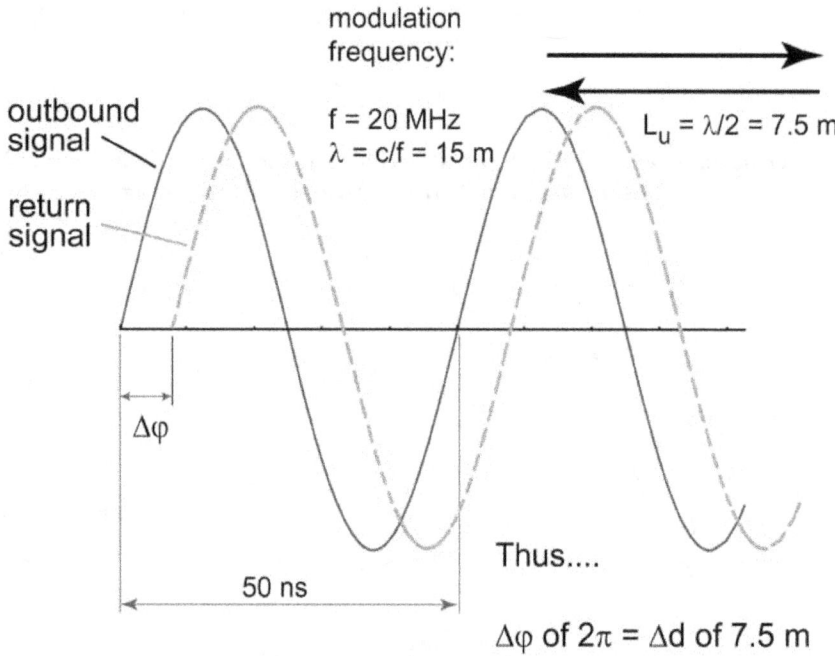

Figure 14. Phase–based measurement of TOF. The unambiguous range Lu is set by the wavelength of the modulation frequency.

For a modulation frequency of 20 MHz, the value for the ambiguity interval is calculated to be 7.5 m. If we assume a phase measurement resolution of 0.01° then range resolution is about 0.25 mm [23].

Higher modulation frequencies will improve the range resolution, but, will also reduce the unambiguous range of the measurement system. One way of maintaining good resolution over large ranges is by using several modulation frequencies (known as heterodyne operation). The lower frequency signal is used to establish the unambiguous interval for the longest measurement distance within which the higher frequency measurement is located. The latter technique has been implemented in imaging products provided by Z+F, Basis Software and others [1].

The general form for calculating the measurement uncertainty for the phase-based approach is [1]:

$$\Delta d = \frac{\lambda \sqrt{2} \sqrt{1 + \frac{K^2}{2}}}{4 \pi \sqrt{N} K (SNR)}$$

where N is the number of equally spaced places the signal is sampled along a given wavelength, λ is the modulation frequency wavelength and K is the modulation contrast which is given by:

$$K = \frac{a_1 a_2}{2}$$

where a_2 is the amplitude of the transmitted modulation signal and a_1 the amplitude of the received signal.

An example of the calculated range uncertainty for a phase based Homodyne 3D imaging system is illustrated in Figure 15 for a variety of modulation contrasts and possible SNR figures. If the available modulation contrast is selected to be 0.1 and the SNR of the detection process is assumed to be 70 dB, then the range uncertainty is about 1.1 mm [1].

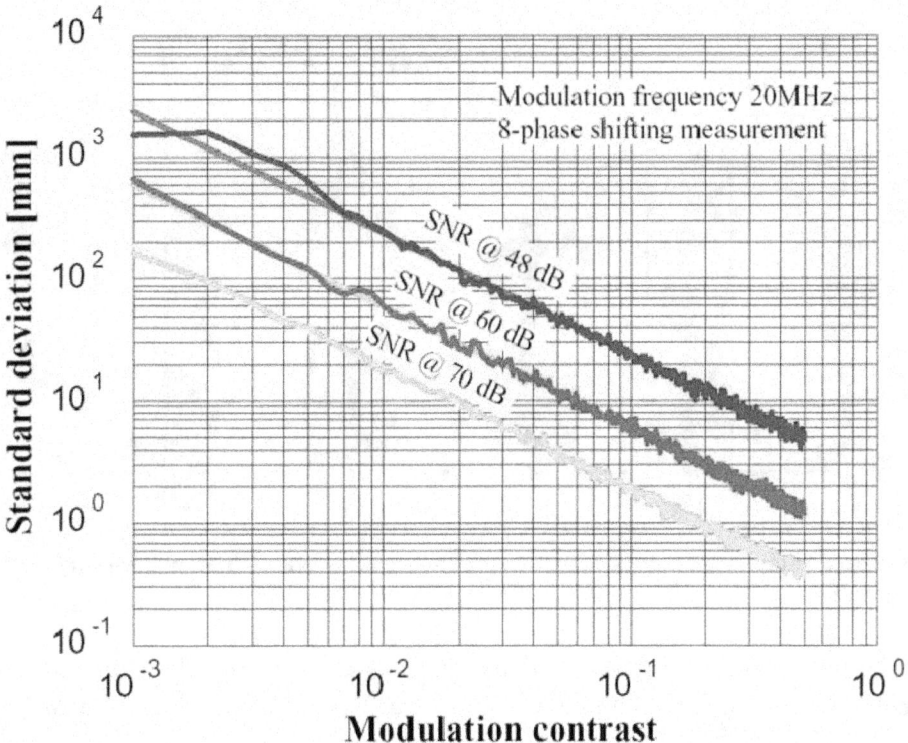

Figure 15. Range uncertainty of an AM-CW Homodyne 3D imaging system based on CMOC photonic mixer technology as a function of modulation contrast (K) and SNR for a single modulation frequency of 20 MHz and a sampling constant, N, of 8 [35].

There are many ways of improving the signal to noise value for a sensor, such as: averaging a large number of samples (often thousands of samples for each measurement); using higher power illumination sources and more efficient optics; higher efficiency photoelectron production in the detector material; and having a digitizing system with enough bits to resolve the higher resolution count. Implementing some of the methods can result in much improved range measurement uncertainty values [1]. Typical range uncertainty values for AM-CW approaches are reported in [4] to be between 0.05 mm to 5 mm for ranges from 1 m to about 100 m.

There are several variations of this approach which should also be discussed. These are:

(a) AM Homodyne 3D imaging

One way of measuring the phase shift between the transmitted and received signals is by synchronously demodulating the signal in the detector through a process known as "photonic mixing". This is a cross correlation process (sampling at specified phase locations of the modulation signal) which allows for the direct derivation of the phase angle. This sampling approach of the mixed transmitted and received signals is illustrated in Figure 16.

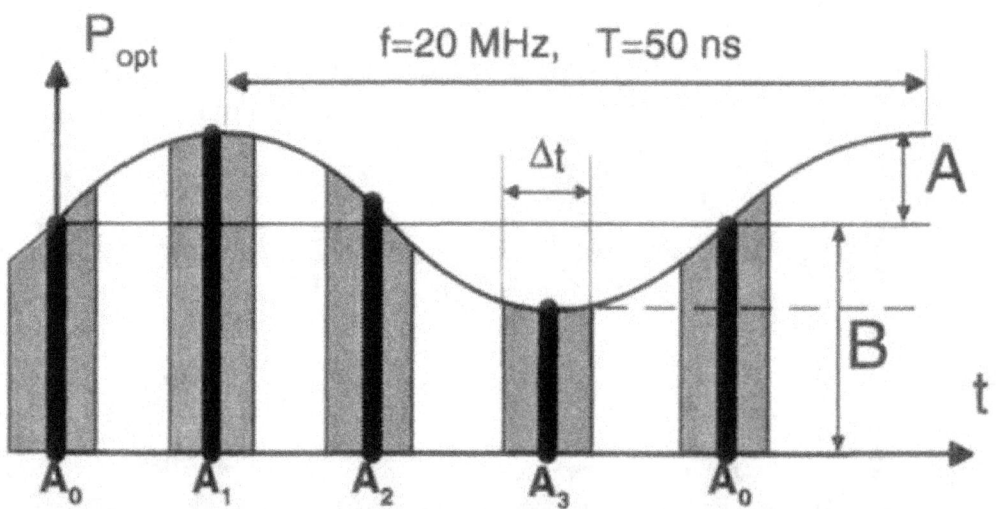

Figure 16. Illustrated "photon mixing" solution where the signal is sampled at an interval of $\pi/2$ along the resulting waveform [7].

Sampling the mixed signal at the detector at intervals of $\pi/2$, or four equally spaced temporal points, and using the discrete Fourier transform, allows for the calculation of the phase delay, signal amplitude, and signal bias. Further in-depth discussion of this approach can be found in [1, 7, 35, 36].

The physical implementation of this measurement approach can take the form of a focal plane array detector. Miniature 3D imagers have been built by CSEM, PMDTec, Canesta and possibly others which take advantage of custom built CMOS/CCD(complementary metal-oxide semiconductor/charged couple device) technology: [www.CSEM.com, www.PMDTec.com, www.canesta.com]. The scene is illuminated using an LED array which matches the FOV of the detector optics. The optical signal is typically modulated at 20 MHz which has an ambiguity range of 7.5 m. The phase delay is measured by each pixel in the detector array. Array sizes of 200 x 200 have been achieved in commercial products. The results achieved with this approach are full frame (array size) range images generated at video frame rates or higher. Claimed values for range resolution and range uncertainty by developers are in the 1 mm range.

The main drawback of this approach has been the inability of the CMOS detector electronics to limit the effects of bright sunlight. This results in a very large value for signal bias value (B), which significantly reduces the signal to noise ratio for the sensor. That is the reason why they have been primarily used in indoor environments. Recently, however, the use of ambient light suppression techniques and laser based illumination are being considered as approaches for overcoming this limitation [www.PMDTec.com].

(b) AM-CW Heterodyne 3D imaging

As explained earlier, one way of maintaining good resolution over large ranges is by using several modulation frequencies (known as heterodyne operation). The lower frequency signal is used to establish the unambiguous interval for the longest measurement distance within which the higher frequency measurement is located. The technique has been implemented in imaging products provided by Z+F, Basis Software and others [1]. An in-depth discussion of this approach can be found in a paper by Schwarte [37]. Although a 0.001 mm range resolution has been reported by one company (under best conditions), typical values are at about 0.1 mm. Typical values of range uncertainty are between 0.5 mm to 5 mm (depending on the range).

(c) Combination of CW and Pulsed Modulation (taken from Lange [7])

"*A combination of both CW modulation and pulsed modulation would be ideal, combining their specific advantages of (1) better optical signal-to-background ratio than available from pure CW-modulation, (2) increased eye safety due to a low mean value of optical power and (3) larger freedom in choosing modulation signals and components. Combination in this content means, for example, a (pulsed sine) operation, i.e. a sequence of 1ms high power sine modulation followed by 9 ms of optical dead time. This dead time can be used for example for post processing tasks*". This technique has been implemented by Trimble in their CX model scanner [product survey at www.gim-international.com]. Range uncertainty is about 2 mm at 50 m distance.

(d) AM-CW Chirp Modulation

In this variation of the traditional AM-CW approach, a linear chirp amplitude modulation signal is used instead of a single frequency. This approach, developed at the Army Research Lab (ARL) by Stann et al. [38], is based on an approach developed for radar [39]. The modulation signal, which looks like the one in Figure 17, is fed into a wideband power amplifier that generates the current drive signal to a

25

semiconductor laser diode. The result is a laser light illumination beam whose intensity is frequency modulated linearly over the chirp bandwidth.

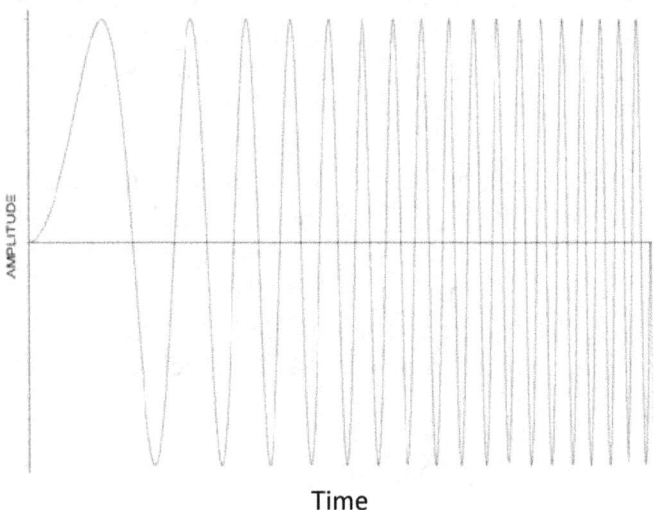

Figure 17. Linear Chirp. Frequency sweep of AM modulation signal (courtesy of ARL).

A custom built self-mixing electronic detector generates a photo-current which is the product of the local oscillator waveform with that of the reflected return signal. This is similar in effect to the photonic mixing technique described under the Homodyne approach. The electronic mixing process results in an intermediate frequency f_{if} being generated which represents the difference between the emitter signal and the reflected signal. The frequency of this signal is proportional to the target range and is equivalent to the measurement of the phase delay in the traditional AM-CW approach. Figure 18 shows the waveforms in the chirped AM-CW modulation approach. The intermediate frequency signal is then processed with a fast Fourier transform (FFT) to obtain the time domain calculation of Δt. Barry Stann describes the details of the approach in [38] and determines that the equation for range resolution is given by:

$$\Delta R = \frac{c}{2\,\Delta F}$$

where ΔF is the difference between the start and stop frequencies of the chirped AM modulation signal. An experimental breadboard was built by ARL where ΔF was equal to 600 MHz. Therefore, the achievable range resolution is equal to about 250 mm.

26

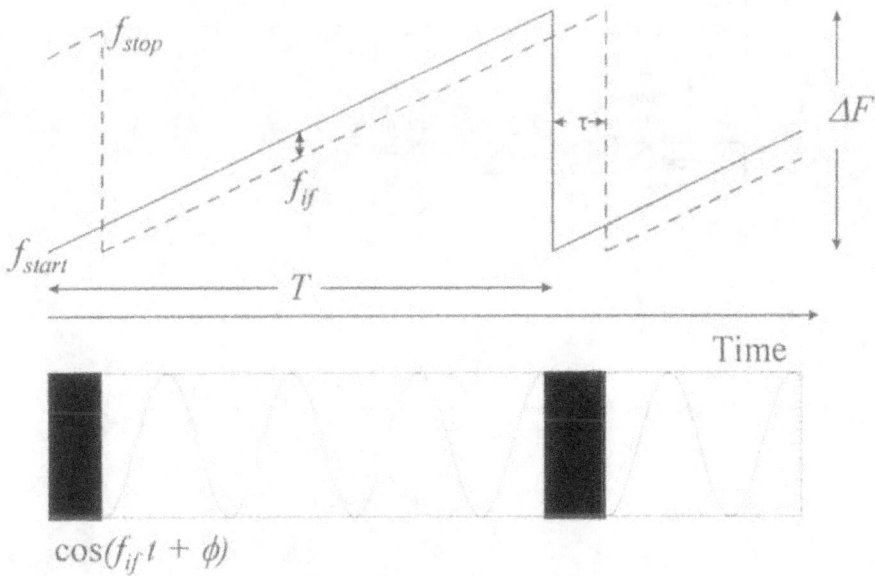

Figure 18. Chirped AM/CW ranging concept. A self-mixing detector is used to measure the product of the local oscillator optical waveform and the reflected return signal – producing the intermediate frequency f_{if} (courtesy of ARL).

Although the initially attained value does not appear to be very high, there are ways to improve range resolution [40]. The main advantage of this approach is that it can be used to recover the time history of multiple target reflections during the time duration of the modulation sweep signal. The AM-CW Chirp approach has generated interest in the commercial sector, however, there are no such products available at this time. The main limitations of implementation include: complicated chirp-generation electronics; limited frequency response of the tunable laser source; and more difficult post processing electronics, especially if a large array implementation is planned.

(3) Coherent Laser Radar or FM-CW Chirp Modulation

In coherent laser radar detection, a part of the illumination laser source and the reflected return signal from an object are optically mixed in a heterodyne detector where the optical interference signal is measured. However, it is more common to have a separate Local Oscillator (LO) and to use mode locking to synchronize the LO to the illumination source. A simplified diagram of the coherent laser radar approach is shown in Figure 19.

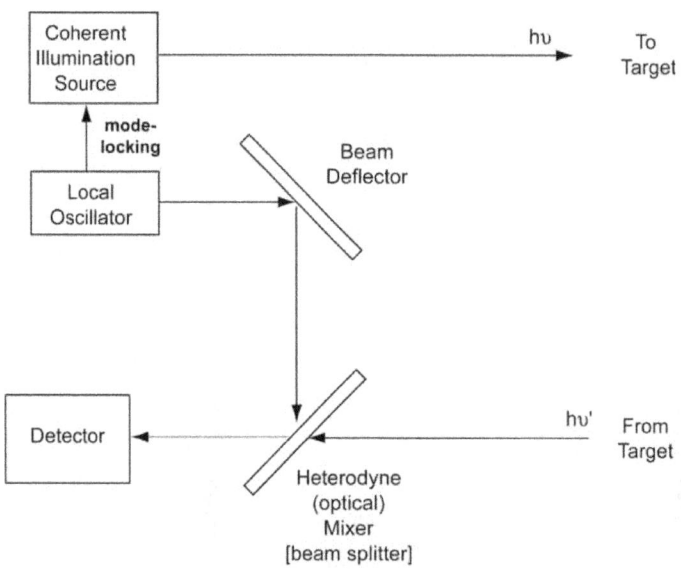

Figure 19. A simplified diagram of the coherent laser radar approach.

In this approach, the LO power being applied to the optical mixer can be adjusted independently to any desirable level (up to the point where it saturates the detector). This feature greatly amplifies the interference portion being detected at the detector. The advantage of this approach is that the signal amplification takes place in the optical domain before any thermal noise is detected in the electronics. This can result in a very high SNR in the detector. A description of the fundamentals of the measurement concept is provided in Chapter 2.4 of the NISTIR 7117 [1]. More detailed information is available from Kammerman [41] and Gatt [42].

A common hardware implementation of this approach is called the FM-CW Chirp laser radar where the wavelength of the laser source is swept linearly using an FM approach. Figure 20 shows the FM chirp approach and the time delay separation between the transmitted signal and the return signal. One way of changing the wavelength of the laser source is through thermal excitation of the laser cavity. Another way involves the use of mechanical piezoelectric actuators to control the length of the laser cavity. This approach requires a very repeatable linear wavelength sweep control system in order to achieve high range resolution and very low range uncertainty values. This requirement is the hardest thing to achieve with this design, however, solutions have been implemented at Nikon Metrology (formerly Metric Vision), Boeing and Bridger. Typical bandwidth of the FM chirp is 100 GHz over a sweep duration of 1 ms. The mathematical validation of the measurement approach is provided in Chapter 2.4 of the NISTIR report [1] along with an implementation scheme provided by Nikon Metrology Inc./Metric Vision. Figure 21 shows a typical implementation of an FM-CW chirp laser radar (courtesy of Nikon Metrology Inc./Metric Vision) Chapter 2.4 of the NISTIR report [1] also provides a technical description of the achievable theoretical range accuracy for this kind of a sensor. Much more detail is provided in [41, 42].

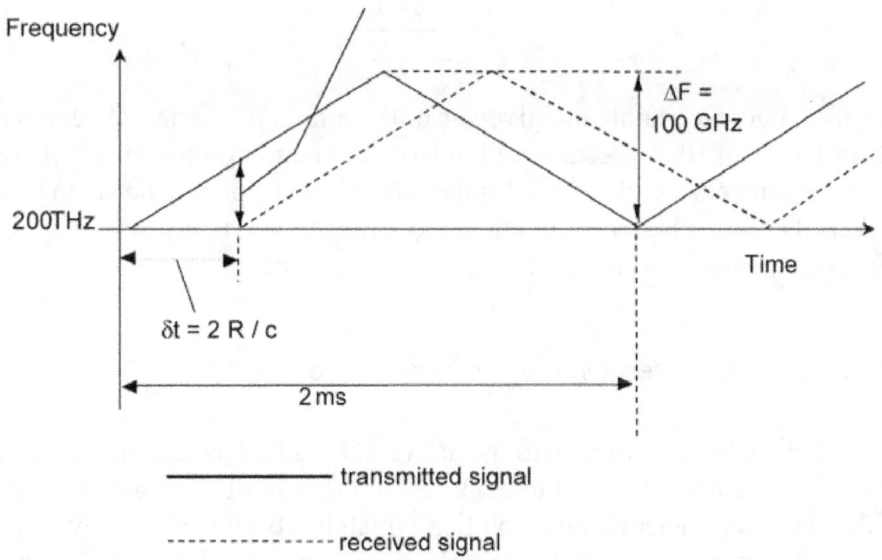

Figure 20. Example of FM-CW chirp concept in measuring time delay between the transmitted and reflected laser signals.

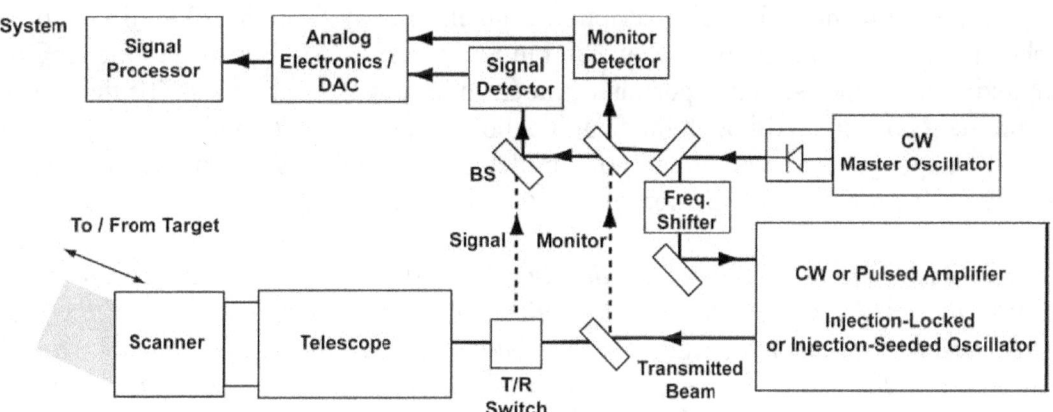

Figure 21. Typical implementation of a FM-CW chirp laser radar (courtesy of Nikon Metrology Inc./Metric Vision).

The equations for calculating range uncertainty values for the FM-CW chirp approach are provided in [4]. As described earlier, the range uncertainty δ_r, for laser based range imaging techniques can be expressed by

$$\delta_r \approx \frac{K}{\sqrt{SNR}}$$

For the FM-CW chirp modulation approach, K is given as

29

$$K = \frac{\sqrt{3}}{2\,\pi}\,\frac{c}{\Delta F}$$

Typical values for range uncertainty are given as 0.010 mm to 0.250 mm for depth of field measurements of less than 10 m. Because of the large time duration for each FM sweep interval (1 ms), range measurements are obtained at rather slow rates (typically between 0.01 kHz to 1 kHz). However, if extreme high resolution and low range uncertainty values are necessary for metrology 3D imaging applications, this approach is hard to beat.

3.2.1.2.3 Interferometric-based Optical Coherence Tomography

As mentioned earlier in the introduction to the active 3D imaging concepts, optical coherence tomography (OCT) has gained considerable interest in the manufacturing applications area. This is because of the sub-micrometer depth resolution which has been achieved by this technique. OCT is basically a low-coherence interferometric (white-light interferometry) technique which uses white (broadband) or near-infrared light sources to generate interference distances in the sub-micrometer range. In addition, the use of near-infrared wavelength light allows the light to penetrate into opaque and transparent materials thus generating high resolution tomograms of the internal structure of a material.

Much of the following technical description is taken directly from the 2006 dissertation by Stephen Beer at the University of Neuchatel in Switzerland[18]. Because of the length and complexity of the thesis, only a portion of the introduction is used to describe the fundamental measurement concept. Although the main goal of the thesis was to realize a very fast and robust 3D OCT imaging system with the use of parallel OCT techniques, it provides a technical overview of the traditional OCT approaches and lists many of the references on this topic.

"OCT and the related technique white-light interferometry for surface detection is based on low-coherence interferometry. A typical interferometric set-up is the Michelson interferometer shown in Figure 7. The beam splitter divides the optical wave from the source into two paths and recombines them after they have been reflected by mirrors. The recombined wave is detected by an optoelectronic device. If the light from the two beams is at least partially coherent, i.e. the phase fluctuations of the two waves are related instead of being completely random, the two waves interfere. This means that the detected optical power depends on the difference of the lengths of the two paths. Typically, the power is sinusoidally modulated as function of the optical path difference with a period corresponding to the wavelength. The interferogram is the measured optical power at the detector as a function of the position of one of the two mirrors called reference mirror. If the illumination source of the interferometer is coherent, e.g. a laser, the signal modulation stretches over a large interferometer arm length difference of up to many meters. If the source is so-called low temporal coherence, such as an light emitting diode (LED) or a thermal lamp, the width of the modulated packet, called coherence length, becomes very small and may even be of the order of a micrometer. If now the reference mirror is moved in a controlled manner and the optical signal is analyzed to detect the signal modulation,

the position of the static mirror can be detected unambiguously with a precision corresponding to half of the coherence length. In OCT, the static mirror is replaced by a sample, and since every reflecting interface of the sample generates a signal modulation at the corresponding reference mirror position, cross-sectional images can be acquired even through scattering material [43, 44].

Most systems exploiting the above explained time-domain (TD-) OCT principle, acquire a depth scan for one lateral position of the sample at a time, a so-called A-scan, then move to a laterally neighboring position to acquire the next scan, and so on. Cross-sectional or volumetric images are then composed of a multitude of A-scans. Real-time volumetric imaging, which is a requirement for many applications, is with this principle only achievable by the use of extremely fast axial scanners. By replacing the single-spot detector by a detector array, a three-dimensional image is acquired by one depth scan[45-47]. This so-called parallel OCT (pOCT) or full-field OCT also reduces the complexity and cost of the systems."[18].

"Recently frequency domain (FD-) OCT principles, alternatively referred to as spectral domain (SD-) OCT, have gained a lot of attention due to their high sensitivity even at low illumination powers. Instead of mechanically scanning the optical delay between the sample and the reference path as in the presented TD-OCT, the optical spectrum of the detected signal is analyzed. The depth profile is then generated by means of Fourier transformation.

Spectrum analysis can be either be done sequentially, e.g. by changing the center wavelength of a monochromatic tunable laser, which is referred to as swept source (SS-) OCT, or simultaneously for example by means of a broadband illumination source, a grating, and a CCD or CMOS line sensor, usually called Fourier-domain OCT.

FD-OCT offers and advantage in sensitivity compared to TD-OCT, especially for low light level applications [48, 49]"[18]. This is discussed in more detail in Chapter 4 of Beer's thesis.

Beer concluded that

"the performance of an OCT system is characterized by the following properties: the number of acquired voxels per complete two-dimensional or three-dimensional image, the axial and lateral resolution, the depth range, the acquisition time, the dynamic range, and the sensitivity, which represents the smallest detectable reflection. These properties are all related and physically limited by the choice of the illumination source" [18].

3.2.2 Research Overview

At the start of the active optical 3D imaging study in 2009, representatives of commercial product developers and research organizations were contacted to determine their interest in contributing information and material on advanced research for the 3D hardware portion of the study. Appendix A contains the names of the organizations contacted. Because of company policies and new government restrictions on releasing information on critical technology to the public, only a limited number of respondents contributed research material for this publication. Therefore, the topics selected represent only a small fraction of the ongoing research. Interested participants contributed information and expert opinions on the needs and solutions for various applications in manufacturing, industrial metrology, automotive safety, the military, dynamic perception, and mobility. These cover many of the needed 3D imaging solutions described in Chapter 4. Some of the presented material is in the form of technical synopses provided by the contributors, while others are technical summaries or text material taken from published documents.

3.2.2.1 <u>Experimental femtosecond coherent LIDAR</u>

The NIST Laboratory in Boulder, Colorado has developed an experimental femtosecond coherent 3D imaging system that has the ability to achieve better than 1 μm range measurement uncertainty even at large-range ambiguity. They also claim the following very desirable features: high precision, speed, low light level operation, multiplexing capability, flexibility and spurious reflection immunity.

It uses two Mode-Locked coherent broadband fiber laser frequency comb sources. One is used for the target signal pulses and the other for the time delayed LO pulses (slightly different repetition rate). The following is a short description of the approach provided by Nathan Newbery at the NIST Boulder office:

> *"The dual comb LIDAR system can be viewed in two different ways – either as a very fine resolution "time-of-flight" system or as a very dense multi-wavelength interferometer.*
> *In terms of the "time-of-flight" picture, the pulses from the signal source are transmitted out and reflected off the target. The second "LO" source then samples the returning LIDAR pulses with very high time resolution. This operational picture is very similar to a down-sampling oscilloscope where the LO pulse provides the sampling time gate that is slowly "walked through" the returning signal. The result is that we map out the electric field of the returning signal pulses. (While standard time-of-flight measurements would just measure the intensity of the returning signal pulse, we measure the full electric field.) By measuring the time of arrival of the pulse envelope, we can obtain a coarse range measurement, with a range ambiguity of 1.5 m set by the laser repetition rate. Since the measurement is coherent, we also measure the phase of the optical carrier wave. From this carrier phase, we obtain a fine range measurement with a range ambiguity given by the optical wavelength, just as with a CW laser interferometer. If the course range measurement is sufficiently precise, we can determine the absolute range to within an optical wavelength (≈1.5 μm). At that point, we can use the interferometric range measurement for finer precision. Figure 22 shows this basic concept and is directly taken from our Nature photonics article"* [50].

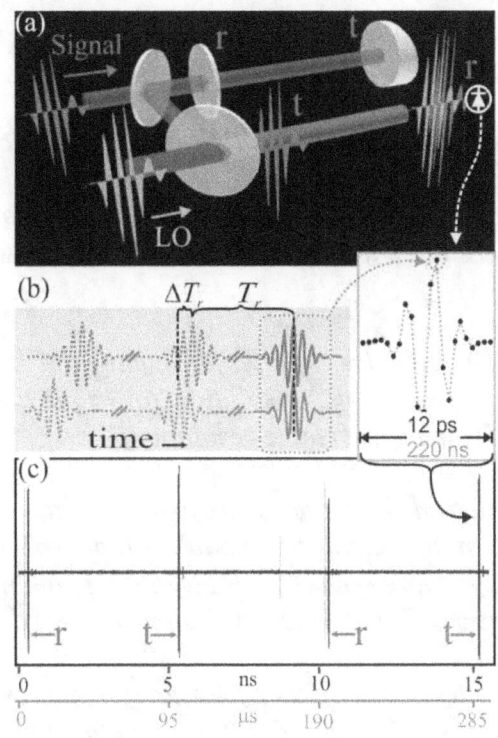

Figure 22. Time-of-flight picture. (a) Ranging concept. A high repetition rate "signal" source transmits pulses that are reflected from two partially reflecting planes (glass plates): the reference (r) and the target (t). The reference is a flat plate and yields two reflections, the first of which is ignored. Distance is measured as the time delay between the reflections from the back surface of the reference flat and the front of the target. (b) The signal pulses are detected though linear optical sampling against a local oscillator (LO). The LO generates pulses at a slightly different repetition rate. Thus, every repetition period (T_r), the LO pulse "slips" by ΔT_r relative to the signal pulses and samples a slightly different portion of the signal. Consecutive samples of the overlap between the signal and LO yield a high-resolution measurement of the returning target and reference pulses. Actual data are shown on the right side where each discrete point corresponds to a single digitized sample and only the immediate overlap region is shown. (c) The measured voltage out of the detector in both real time (lower scale) and effective time (upper scale) for a target and reference plane separated by 0.76 m. A full "scan" of the LO pulse across the signal pulse is accomplished every $\approx$100 µs in real time and every $\approx$ 10 ns in effective time. Two such scans are shown to illustrate the fast, repetitive nature of the measurement. Also seen are two peaks in grey which are spurious reflections of the launch optics.

"As stated above, the system can also be viewed as a very dense multi-wavelength interferometer. The coherent pulsed sources are really frequency combs – they output an entire comb of different wavelengths of light at once. Through the use of the second comb, we essentially configure the system so that each one of these wavelengths of light is used in its own individual CW laser interferometer and provides its own range measurement. This system then is similar to a conventional multi-wavelength interferometer (MWI) except that we have many, many (thousands) of different wavelengths involved. Through the standard MWI approach of combining the range measurements from all these individual CW interferometers, we acquire an absolute range measurement with high precision. The use of the combs, as opposed to conventional swept CW lasers, allows us to do this quickly, with only two laser sources, and with very low systematic errors." [50].

Performance advantages of the femtosecond 3D imaging system are provided in comparison to the FM modulation (chirp approach) coherent laser radar approach:

"The resolution of any laser system is set by the system bandwidth divided by the signal-to-noise ratio. The dual comb system can have significantly more bandwidth than conventional FM modulated laser radar and therefore can provide significantly higher range resolution. There are three advantages to the dual comb system vs. FM modulated coherent laser radar:

1) Resolution: Much higher effective modulation bandwidth is possible (THz vs. GHz) with a corresponding improved range resolution (or, equivalently, reduced signal-to-noise ratio requirements for the same range resolution).

2) Speed: Multiple modulation frequencies are transmitted at once for this system as opposed to sequentially in a typical fm modulated approach. As a result, we can determine an absolute range more quickly in a given range ambiguity window. Sampling rates of at least 5000 measurements per second can be achieved.

3) Accuracy (low systematics): Significantly lower systematics are possible since there are no cyclic errors. The reason for this is that the simultaneous transmission of multiple modulation frequencies allows us to observe and "gate out" spurious reflections that would otherwise lead to cyclic errors, a major systematic in other ranging systems. We note that this control of systematics is hard to quantify but is potentially one of the most beneficial features of the system." [50].

This system is still a lab research instrument and has not been implemented with a scanner. However, if implemented, the features offered by this system would make it a high performance laser-based 3D imaging system that could be used for large-scale metrology applications. Issues of sensor complexity and cost would have to be resolved in order for this approach to be developed into a useful and cost competitive 3D imaging product for industrial and other applications.

3.2.2.2 Active laser stabilization during linear sweeps for FMCW chirped LADAR

This technical synopsis was provided by Brant Kaylor of Bridger Photonics and is reproduced with his permission.

Bridger Photonics (Bridger) has developed a new design for a frequency-modulated continuous-wave (FMCW) LADAR system that utilizes an actively stabilized swept laser source and a heterodyne receiver architecture similar to that of microwave radar systems. The system offers tremendous optical bandwidth and extremely linear chirps, thus producing extremely high range resolution and very low measurement noise.

"A detailed diagram of the FMCW technique is shown in Figure 23. Bridger's innovation enables extremely linear chirps over enormous optical bandwidths to produce

unprecedented resolution and precision. The frequency chirped laser light passes through a splitter, with the transmit portion passing first through an optical circulator and then towards a distant target object. Light that returns from the target is time delayed by τ_D= 2R/c, where R is the distance to the object and c is the speed of light. This time-delayed light is collected, passes out of the circulator to be recombined with the portion of the original chirped light that is not time delayed called the reference or local oscillator. When the recombined light is detected, a constant frequency offset exists between the two chirps as a result of the time delay. This appears as a heterodyne or "beat note" which has a frequency f_{beat} = $\kappa\tau_D$ = 2κR/c, where κ is the chirp rate in Hz/s. The distance to the object, R, can therefore be determined by measuring the beat note frequency. As before, the range resolution ΔR of this LADAR technique is given by ΔR = c/2B, where B is the bandwidth of the optical chirp. Typically, FMCW chirped LADARs have much better range resolution because the optical bandwidths that they can cover are much larger. The maximum range window or beat note that can be measured is often limited by the bandwidth of the digitizer used and can reach large distances (> 1 km). Note that by delaying the reference path, the range window can be shifted, so that an object could be examined tens of kilometers away."

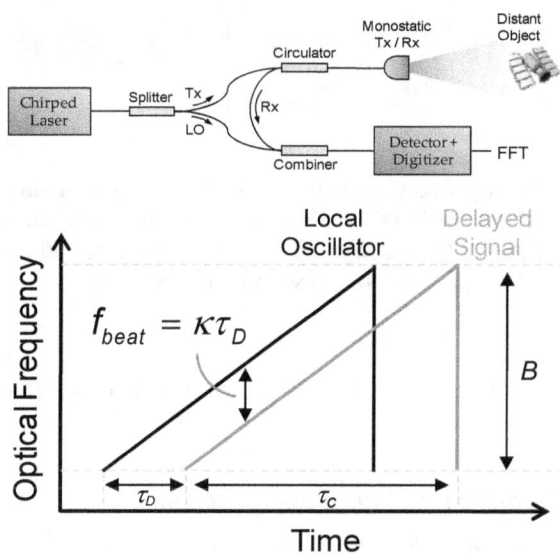

Figure 23. (Top) The setup for an FMCW chirped heterodyne LADAR system. (Bottom) The optical frequency versus the sweep time for the local oscillator and delayed return signal.

"To date, Bridger has achieved range resolutions of 47 μm and precisions of 86 nm. These results utilized an optical sweep linearized over 5 THz with frequency residuals below 200 kHz (4 parts in 10^7) [51]. These results are shown in Figure 24. In the interest of clarity, Bridger measures resolution as the full-width half-maximum of the range peak for a single-point return and precision as the standard deviation of the least-squares peak fit."

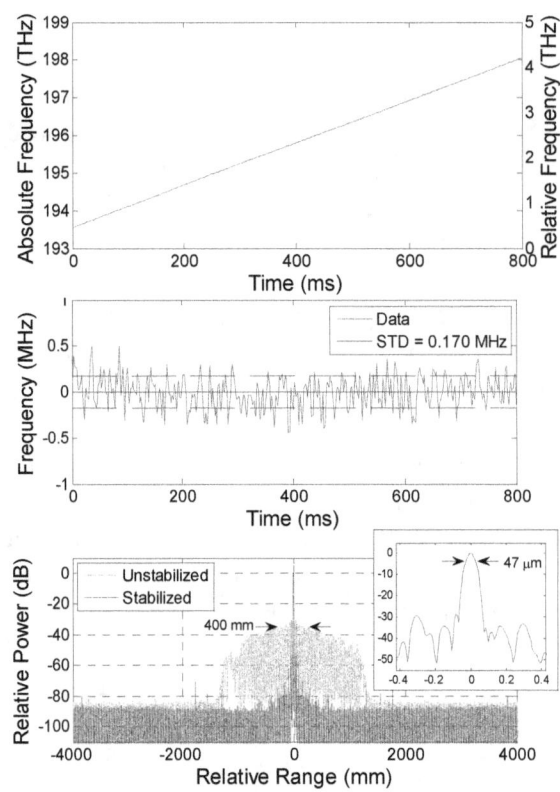

Figure 24. (Top) A frequency sweep from the SLM-H laser system shown covering more than 4 THz of optical bandwidth. (Middle) The residual sweep errors from linearity showing a 170 kHz standard deviation. (Bottom) Two range profiles for (grey) an unstabilized sweep with 400 mm FWHM and (black) a stabilized sweep with 47 μm FWHM.

The following is a list of some of the main advantages offered by the new FMCW laser radar system design:

1. *Extremely high range resolution and precision.*
2. *Sensitivity to very low light levels. Bridger's system relies on heterodyne detection and thus is sensitive to very low light return levels. Bridger has demonstrated a unity SNR point of less than 100 aW (or 1 x 10⁻¹⁶ W).*
3. *Low bandwidth receiver electronics. FMCW systems rely on heterodyne detection to measure the frequency offset between the return light and local oscillator. This frequency offset is typically on the order of (1 to 10) MHz, significantly less than the bandwidth of the optical sweep.*
4. *Flexible resolution and measurement speed. Bridger Photonics has developed a variety of systems to offer resolution from 50 μm to 5 mm at update rates of 1 Hz to 30 kHz. At present, the Bridger system can achieve precisions < 100 nm at 10 Hz and < 1 μm at 100 Hz.*
5. *Large range windows and stand-off distances. Through active stabilization, Bridger has produced very linear chirps with long coherence lengths. In practice, the analog-to-*

36

digital converter is often the limiting factor. To date, Bridger has demonstrated 100 m range windows and up to 14 km stand-off distances.

6. ***Robust, compact, low cost, commercial off-the-shelf components.*** *Bridger leverages the extensive development of components in the telecom band (1.5 µm) to enable robust, compact, low cost systems. Existing systems are housed in standard 483 mm (19 in.) rack-mounted enclosures.*

3.2.2.3 <u>Non-scanning parallel optical coherence tomography (pOCT)</u>

CSEM in Switzerland has developed a non-scanning parallel OCT – white light interferometer sensor which is being marketed by Heliotis (http://www.heliotis.ch). This is part of the EU project SMARTIEHS (http://www.ict-smartiehs.eu:8080/SMARTIEHS/publications) [52] to develop a massively parallelized interferometry based OCT system for testing and characterizing of MOEMS (Micro- Opto-Electro Mechanical Systems) structures. CSEM is in charge of developing a pOCT 3D imaging system which has a spatial array size of 150 x 150 pixels. This action was taken because the current state-of-the-art inspections systems perform serial inspection of each MOEMS structure individually. This procedure is too time consuming and is not suited for mass production needs. SMARTIEHS is expected to decrease inspection time of wafers by a factor of 100 which will cut production costs and shorten the time-to-market. The following material on the new massively parallel inspection approach comes from the chapter on Photonics from the CSEM 2008 Scientific and Technical Report (http://www.CSEM.ch).

"A new inspection approach has been developed in the EU project SMARTHIES [52]. With the introduction of a wafer-to-wafer inspection concept, parallel testing and characterization of tens of MOEMS structures within one measurement cycle has been made possible. Exchangeable micro-optical probing wafers are developed and aligned with the MOEMS-wafer-under-test. Two different probing configurations, a 5 x 5 array of low-coherence interferometers (LCI) and a 5 x 5 array of laser interferometers (LI), use an array of 5 x 5 smart-pixel cameras featuring optical lock-in detection at the pixel level [53].

This smart-pixel image sensor dedicated to optical characterization of MOEMS, called SMOC, offers processing of the detected intensity signal at each pixel. In an interferometer, this functionality enables to pre-process the interference signal in order to increase measurement speed and accuracy. The SMOC imager, with its 150X150 pixels, features background suppression and direct I-Q (smart pixel) demodulation [18]. Global shutter allows integrate-while-read operation. The sensor has column-parallel 10-bit ADCs and 5 output pads operating at 41.5 MHz each. Maximum frame rate of more than 400 fps (frames per second) at full resolution is supported. Higher frame rates can be achieved by region-of-interest (ROI) sub-frame viewing. On the digital side, a fully programmable sequencer has been implemented. The camera composed of the 5 x 5 synchronously operating smart-pixel imagers has a CameraLink-interface running at 3 Gbps.

The principle of the camera module allowing individual mechanical adjustment of each imager module is shown in Figure 25.

A prototype of the smart-pixel imager is under development and will be taped out at the beginning of 2009. Further versions will have quadrupled number of pixels (300 x 300) and a significantly increased frame rate (more than 9 kfps)."

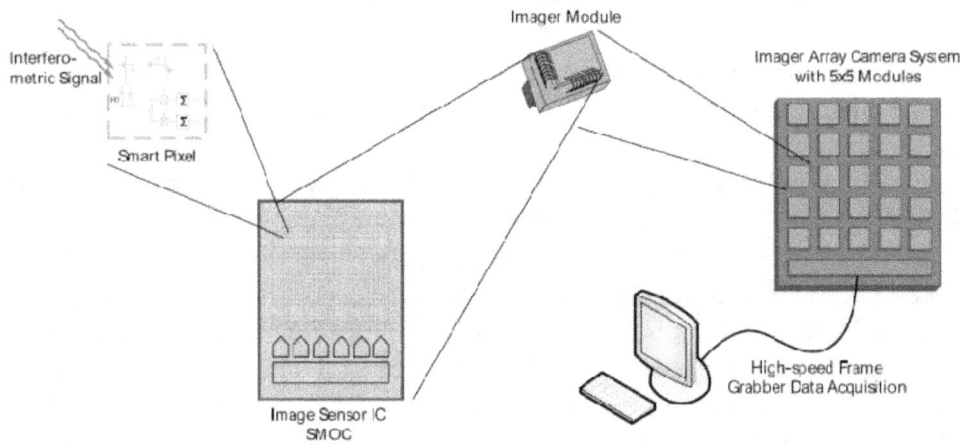

Figure 25. Schematic representation of the SMARTHIES camera module - from the pixel to the high-speed interface to the PC. (from SMARTIEHS website link [52] – Newsletter N.2).

In a report issued in April of 2010, SMARTIEHS reported that the design of the massively parallel inspection system has been completed and that two single channel demodulation detectors have been tested. The multi channel system is expected to be completed in 2010. Further information can be obtained directly from the SMARTIEHS web site:
http://www.ict-smartiehs.eu:8080/SMARTIEHS/publications

As mentioned earlier, Heliotis has developed several fast 3D microscopy products based on the pOCT concept developed at CSEM. In addition to inspection of micro-optical components (such as MOEMS), pOCT products are also being used in biological tissue and other medical imaging, for inspection of surface mounted devices and die and wire bonds, and inspection of injection molded parts. This represents just a sampling of potential applications for this fast growing technology.

3.2.2.4 Real-time active 3D imaging for dynamic perception applications

An area of active optical 3D sensor research and product development that has demonstrated rapid growth in the last 10 years is in real-time 3D imaging for mobility or dynamic perception applications. Focal point arrays, multi-laser/multi-planar scanning sensors and now Micro-Electro-Mechanical Systems (MEMS) based laser scanning sensors are being built by developers to demonstrate their use for improving perception in autonomous and semi-autonomous vehicles for navigation/collision avoidance, non-contact safety systems, automated material handling, and other industrial, commercial, transportation and military applications. The interest in non-scanning FPA technology has grown quickly and will be covered first.

The FPA designs use either pulsed laser or AM modulated homodyne techniques for range measurement. There are no moving parts used for laser scanning and only a single laser pulse or AM modulated sampling period is used to measure the range image of the FOV at each pixel of the APD array. Although 256 x 256 APD arrays have been built for pulsed laser mode operation, the most common size currently available as a product is 128 x 128. Since the APD approach seems to be best suited for narrow field-of-view applications, some developers are placing these sensors on precision pointing stages (for stitching scenes together) or scanning the detector FOV to generate larger FOV data sets.

A. Custom CMOS/CCD FPA Technology

Several developers have produced compact, low cost, real-time 3D imaging systems that take advantage of custom CMOS/CCD technology and employ the AMCW Homodyne range imaging approach. They are stand-alone camera systems which typically have used LED light emitting array (in the IR signal range) continuous wave signal illuminators which are modulated in amplitude. The reflected signal from the scene travels back to the camera, where the TOF is measured by recording the phase delay between the signals. The signal phase is detected by using a synchronous demodulating scheme. The largest array that is available is a 200 x 200 FPA and frame rates of 100 Hz have been achieved. Performance outdoors was poor in some earlier versions, but developers are designing laser-based illuminators or are developing ambient compensation circuitry at the pixel level to overcome these limitations. Because of their low cost, these types of systems are being increasingly used in automotive, robotic (both military and industrial) applications.

Two of the leading developers of custom CMOS/CCD 3D imaging cameras, PMDTec and CSEM, contributed material for the research part of the active 3D imaging technology study.

(1) PMD Technologies (in Siegen, Germany) and their partners/shareholders Audi and ifm have mass produced and sold over 100 000 sensor units and have generated over 25 patents. They were first to introduce a 204 x 204 pixel array camera and are now exploring the production and qualification of a non-scanning imager for automotive safety applications. The automotive sensor is expected to provide real-time object and pedestrian detection and tracking to a measurement range of about 60 m. The unambiguous range of the sensor is set to 120 m by lowering the AM modulation frequency. This is the maximum possible measurement range if sufficient illumination

power is used. The B-Muster PMD camera has sufficient optical power to obtain range measurements out to 60 m from vertical surfaces. The sensor also uses patented suppression of background illumination (SBI) circuitry in each pixel in order to allow the sensor to operate in bright sunlight conditions. The FOV of the sensor is expected to be about 50° horizontally, however, if a larger FOV is desired, multi-camera operation can be achieved using different frequency channels. Figure 26 shows how the sensor and illumination sources can be mounted on a vehicle.

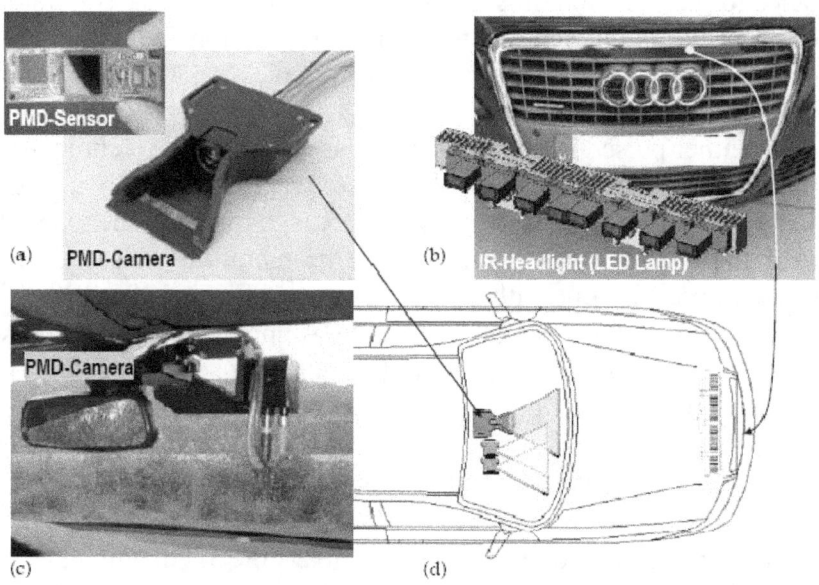

Figure 26. Illustration of a PMD camera mounted on a vehicle: (a) PMD Muster camera, (b) IR LED illumination lamps, (c) Internal car setup, (d) Schematic of car setup.

In 2010, PMD introduced the PMD (vision) CamCube 3.0 camera. With a 200 x 200 pixel array, the sensor is the highest spatial resolution all-solid-state TOF 3D camera that is currently commercially available.

Research activities at PMDtec include: building progressively larger and larger pixel array sizes with smaller and smaller CMOS technologies, increasing frame rates to over 100 fps, providing onboard camera CPU/DSP (digital signal processing) capabilities to increasingly enhance onboard intelligent 3D image processing for advanced data visualization and intelligent control systems.

(2) CSEM and their marketing partner MESA in Zurich, Switzerland also develop and produce AM-CW Homodyne FPA 3D imaging cameras. They use CMOS technology to miniaturize and to reduce cost and power usage in their product. The MESA 3D FPA imager, the SR4000, has an array size of 176 x 144 pixels and can be operated at a frame rate of as high as 54 fps. Although the application areas are similar to those being addressed by PMD, their main focus is on industrial applications, and they have made a tremendous effort to provide stable and repeatable range imaging measurements in very demanding shop floor environments. Even though they have implemented in-pixel

background light suppression capability, this is not the main emphasis of the product design. Therefore, the SR4000 is not intended for use in bright sunlight.

CSEM has been in the research and development business of active optical 3D imagers for close to ten years and was first to successfully demonstrate the capabilities of the lock-in based concept in 3D imagers at the pixel level. They were also one of the first companies to develop and demonstrate the parallel optical low coherence tomography (pOCT) concept. In recent years, they have concentrated their research effort to further miniaturizing the TOF cameras while improving range measurement performance and lowering power usage. The SR4000, marketed by MESA is only (4 x 4 x 4) cm^3 in size, requires less than 6 W of power, and has a range measurement uncertainty of less that ± 1 cm. Figure 27 shows a picture of the camera and Figure 28 shows 3D range images taken with the camera. Higher modulation frequencies (up to 80 MHz) can be used to further improve the range resolution. The research base at CSEM is very broad and includes:

- Ultra low power – single chip digital cameras (2 mW)
- High speed – 2D imaging (1000 fps to 2000 fps)
- Ultra high speed – line sensors (80 000 fps)
- Very high dynamic range miniature video cameras (> 140 db)
- High resolution optical encoders (100 nm)
- X-ray imaging systems
- Phase contrast X-ray imagers

Additional information covering research to extend capabilities of optical TOF 3D imaging can be found in [54].

Figure 27. Picture of the MESA SR4000 3D imaging camera.

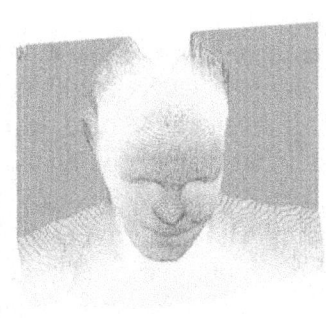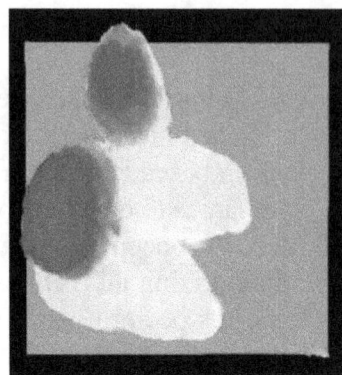

Figure 28. Single frame range images taken with the MESA SR4000 3D camera.

(3) Canesta, in Sunnyvale, California, is also developing CMOS-based FPA 3D cameras that perform like the cameras available from PMDTec and CSEM. Similarly, they also use the continuous wave AM phase-based approach for measuring the TOF at each pixel in the detector array. Custom 3D imaging detectors are provided to existing customers for specific applications and embedded systems, but are also provided to academic organizations for research purposes. The detectors are 64 x 64 pixel arrays than can be implemented with background illumination suppression circuitry and with two different AM modulation frequencies in order to increase the unambiguous range of operation (http://www.canesta.com).

B. Avalanche Photodiode Array (APD) FPA 3D Imagers

(1) Another FPA imaging concept which has matured in the last five years and is being introduced to the sensors market by Spectrolab/Boeing (http://www.spectrolab.com) is the Geiger-mode APD array. This imaging technology was originally developed and demonstrated by MIT Lincoln Labs. Under a technology development agreement, Spectrolab has taken over the research work and is making improvements to the design to make it suitable for volume production, reduced cost, and improved stable and reliable performance. The following description of Geiger-mode operation comes from Section 4.1 of Chapter 7 in the NIST book publication on Intelligent Vehicle Systems [55]:

"In order to achieve enhanced ionization which is responsive to the arrival of a single photon, Lincoln Labs have developed a "Geiger-mode" (GM) avalanche photodiode (APD) array that is integrated with fast CMOS time-to-digital converter circuits at each pixel [56]. When a photon is detected there is an explosive growth of current over a period of tens of picoseconds. Essentially, the APD saturates while providing a gain of typically > 10^8. This effect is achieved by reverse-biasing the APD above the breakdown voltage using a power supply that can source unlimited current."

42

Some of the early work by Lincoln Labs, using a GM APD 32 x 32 FPA LADAR can be found in SPIE conference publications [28, 57].

"MIT also reports [29] that they have extended their earlier work with silicon based APDs by developing arrays of InGaAsP/InP APDs, which are efficient detectors for near-IR radiation at 1.06 μm. These detectors are 32 x 32 pixel arrays, with 100 μm pitches. Figure 29 shows the key elements that are integrated into a package for Geiger-mode operation. In the figure, light enters the array from the top and is focused by the microlens array onto the detector array. About 70 % to 80 % of the light is captured and focused onto the detector pixel elements. The overall detector efficiency with the microlenses is in the order of 30 % to 35 % thus considerably increasing detector efficiency. The thermoelectric cooler is necessary to reduce the dark current rates in order to keep the LADAR time-of-flight gates open for times in excess of 1 μs. This is adequate for many LADAR applications."

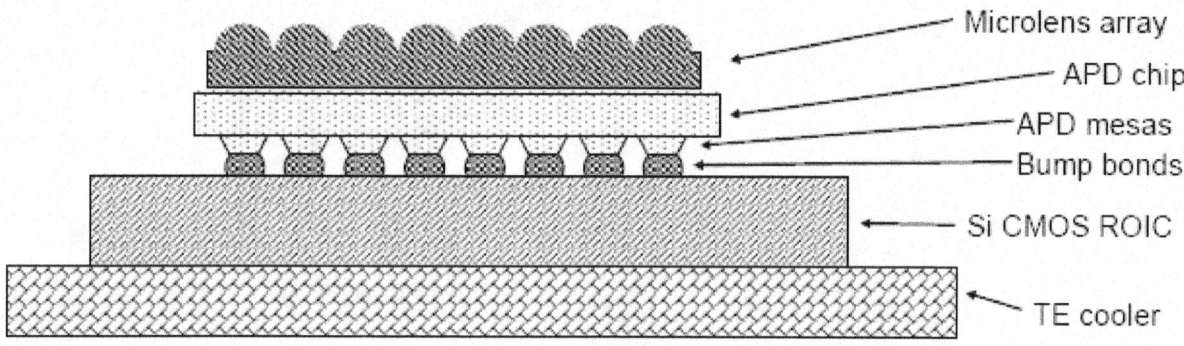

Figure 29. Cross section showing key elements integrated into package for Geiger-mode APD arrays (courtesy of MIT/LL).

Additional details of the Geiger-mode APD approach can be found in Chapter 5.3 of the NISTIR 7117 report [1]. It describes the differences between p-n photodiode detectors, linear-mode APD detectors and Geiger-mode APD detectors. The chapter also describes the active quenching circuitry needed for resetting the GM APD below the breakdown voltage in order to quickly prepare it for taking the next TOF measurement. Much of the material for describing the GM APD operating principles came from Aull et al. [58].

Spectrolab has incorporated improvements in the Geiger-mode pixel design to provide more uniformity in performance across all the pixels in the array, has increased the sensor frame rate to close to 30 fps, and has improved the Read-Out Integrated Circuit (ROIC) to operate with a 0.5 ns timing resolution [30]. Table 3 lists the most important FPA figures of merit that impact the single photon counting sensor system performance. PDE stands for photon detection efficiency and DCR for dark current count rate in the table. Figure 30 demonstrates the operation of a GM-APD. Figure 31 shows a color coded 3D image taken with a Spectrolab Gen I 3D imaging camera. The right image shows the same scene taken with a 2D vision camera.

Table 3. Typical single photon counting FPA figures-of-merit for a 32 x 32 GM-APD (courtesy of Spectrolab)

Parameter	Spec	Test Condition
Wavelength	1.06 µm	
Format	32 x 32	
Pixel Pitch	100 µm	
PDE	40 %	240 K, 4 V overbias
Fill Factor	70 %	
DCR	20 kHz	240 K, 4 V overbias
Cross talk	< 1 %	Optimized overbias for SCS
Pixel operability	99 %	
RMS timing jitter	0.5 ns	

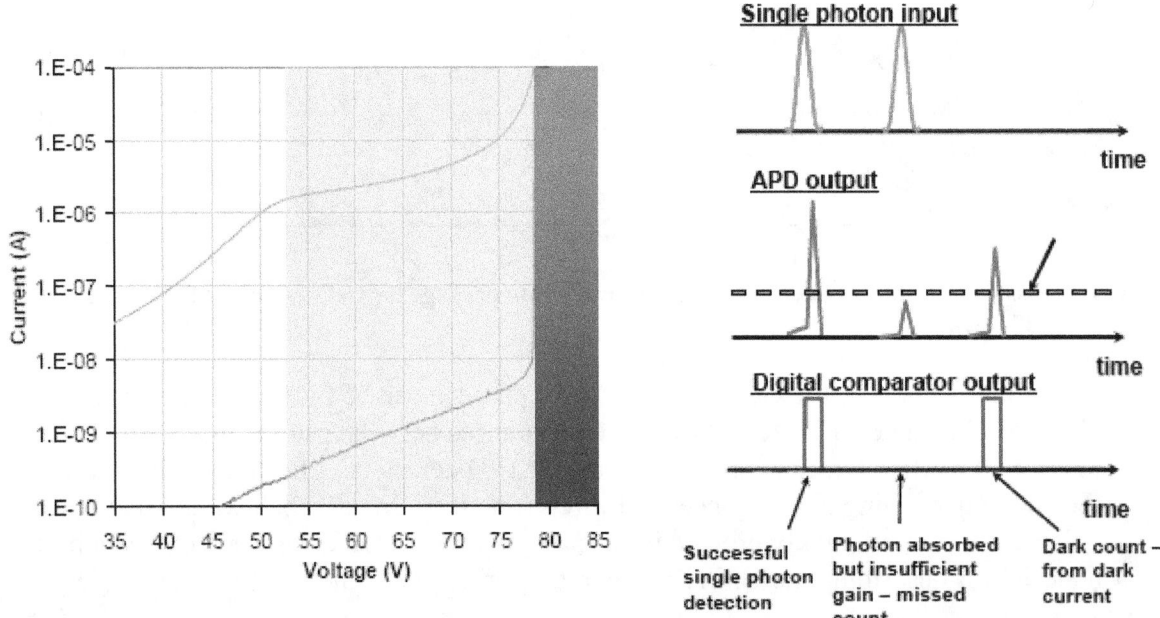

Figure 30. Typical operation of a Geiger-mode APD. The left panel shows the voltage and current characteristics of a 200 µm GM-APD under illumination (red line) and in darkness (blue line). The linear mode region is highlighted in blue, while the Geiger-mode operation region is shaded in red. The right panel illustrates the three possible cases in Geiger-mode detection. (courtesy of Spectrolab).

Figure 31. 3D image taken with a Spectrolab Gen-I 3D imaging camera. The picture is color coded with the range information. The right image is the same scene taken with a 2D vision camera (courtesy of Spectrolab).

The new ROIC in the Spectrolab Gen II LADAR camera is being redesigned in 0.18 μm CMOS technology by Black Forest Engineering. The main effort is in minimizing timing jitter, improving pixel functionality and in reducing crosstalk between pixel elements. With support from DARPA, larger format FPAs (128 x 32 and 256 x 256) can be expected. The main goal of the development work appears to be directed at making the Geiger-mode FPA sensors more efficient, more robust and convenient to operate, and to be commercially competitive.

(2) Advanced Scientific Concepts Inc. (ASC)

ASC, based in Santa Barbara, CA, was the first company worldwide to develop and demonstrate a real-time APD array flash LADAR 3D camera. The latest camera, which uses a 128 x 128 APD pixel array detector, has been successfully demonstrated in various military and commercial applications – including use in space by NASA. In 2006, ASC provided a synopsis of the APD flash LADAR concept for the NIST book publication "Intelligent Vehicle Systems: A 4D/RCS Approach" [55]. The following is part of that synopsis:

"The ASC designs used Commercially Off The Shelf (COTS) parts as much as possible to reduce cost and a compact laser was used to reduce volume, weight and power. The designs used no mechanically moving parts for laser scanning and only a single laser pulse was needed to capture the entire FOV with a hybrid 3D FPA. Figure 32 shows a typical ASC hybrid 3D FPA configuration. It shows how a Readout Integrated Circuit (ROIC) is bump bonded to a solid-state detector array (such as a Silicon, InGaAs or HgCdTe PIN or APD detectors).Each unit cell, or pixel, contains circuitry which independently counts time from the emission of a laser pulse from the camera, to the detection of the reflected pulse from a surface. In this manner, each pixel captures its independent 3D information in a scene (angle, angle,

range). Additional circuitry is available in each pixel to capture temporal information from the returning pulses. Twenty sampling circuits are provided in the ASC FPA design which helps in detecting objects which are obscured by smoke, foliage, etc. As designed with off-the-shelf optics, all the 3D imaging systems use two aperture systems whether they are WFOV (wide FOV), NFOV (narrow FOV) or a combination of the two: an aperture for the laser-transmit optics and an aperture for the 3D imaging, receive optics. Figure 31 illustrates a possible camera configuration for the ASC pulse TOF WFOV design [34]. The drive and output electronic circuit boards, as well as the laser transmitter, are inside the camera housing. The same configuration and the same 3D FPA (with longer focal length optics) can also meet the needs for a NFOV stand-alone sensor." [55]

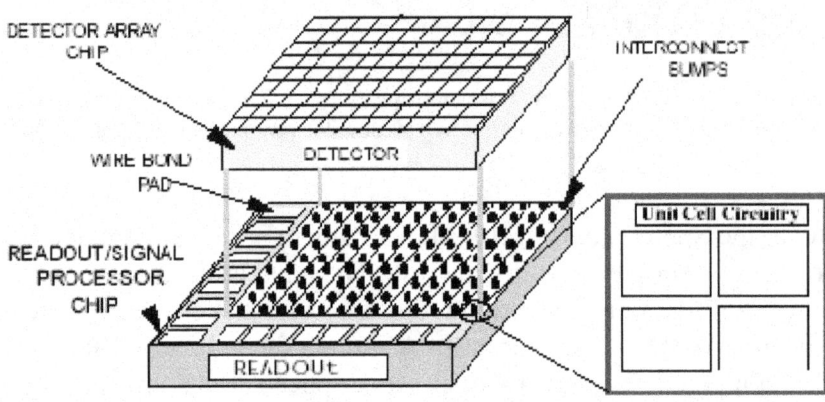

Figure 32. ASC 3D FPA Hybrid Design. ROIC bump bonded to detector array chip.

Figure 33. Possible packaged configuration for a standalone WFOV camera. Estimated weight is 1.8 kg; the COTs optics are a large fraction of the weight.

ASC is now using a 128 x 128 InGaAs APD detector which operates with an eye safe 1570 nm pulsed laser. The new APD detector has achieved an eight times improvement in gain over the original InGaAs PIN detector array. This has allowed them to operate the sensor at longer ranges (over 2000 m) at 20 fps to 30 fps. They have enhanced object detection, recognition and tracking abilities by adding and registering color and/or IR detector images with the range image data taken with the camera. The specifications for the latest APD FPA cameras available from ASC are provided in Appendix B (the technical survey portion) of this report.

The following updated overview on the 3D APD array Flash Camera operation came from the ASC Inc. website (http://www.ASC3D.com).

"ASC's 3D cameras are eye-safe Flash laser radar (3D Flash LIDAR) camera systems that operate much like 2D digital cameras, but measure the range to all objects in its field of view (the "scene") with a single "flash" laser pulse (per frame). The technology can be used in full sunlight or the darkness of night. 2D Video cameras are able to capture video streams that are measured in frames per second (fps), typically between 1 fps and 30 fps with 1, 5, 15 and 30 being most common. The dynamic frame capture paradigm holds true for the 3D Flash LIDAR Camera (3D FLC) as well with 1 fps to 30 fps (or faster if required) being possible. There are some cooling design constraints when designing the laser system for the higher fps capture rates to ensure the laser has adequate cooling operating margins.

As seen in Figure 34, light from the pulsed laser illuminates the scene in front of the camera lens, focusing the image onto the 3D sensor focal plane array, which outputs data as a cloud of points (3D pixels). Each pixel in the sensor contains a counter that measures the elapsed time between the laser light pulse and the subsequent arrival of the reflected laser light onto each pixel of the focal plane. Because the speed of light (the laser pulse) is a known constant, accurately "capturing" the scene in front of a camera is a relatively straight-forward process. There is an inherent relationship between the pixels themselves in the scene, representing the entire scene at an instant in time. The point cloud data accurately represents the scene allowing the user to zoom into the 3D point cloud scene without distortion.

The 3D Flash LIDAR cameras are single units which include a camera chassis or body, a receiving lens, a focal plane array and supporting electronics. On the Portable 3D camera, data processing and camera control is done on a separate laptop computer. The TigerEye 3D camera is controlled via an Ethernet connection with the initial processing done on the camera and display done on a PC. The output of both cameras can be stored and displayed on a PC running ASC's software; TigerView™ for the TigerEye and FLASH3D™ for the Portable camera.

Both raw and processed data can be stored or output in various formats for additional video processing using industry standard 3D computer graphics tools such as Autodesk's Maya, 3D Studio Max or Softimage.

Figure 35 illustrates raw data capture (color coded for range and viewing) of a single-pulse 3D image taken with a Portable 3D FLVC (Flash LADAR Video Camera). Note the FLVC camera can be used to accurately identify vehicles or other objects without additional data. It is possible to "see" through the windshield to identify passengers and objects inside the vehicle as well. In this example, raw data was processed using the ASC's range algorithm (essentially a raw-data image) only, color-coded for depth and intensity.

Note the shape of the rotating helicopter rotor blade, in Figure 36, as it is captured without motion distortion from above. The lack of motion distortion is a major feature of ASC's 3D cameras. The image is color coded for range and intensity which is determined using the ASC's algorithm. The picture on the right is the same raw data rotated for viewing purposes."

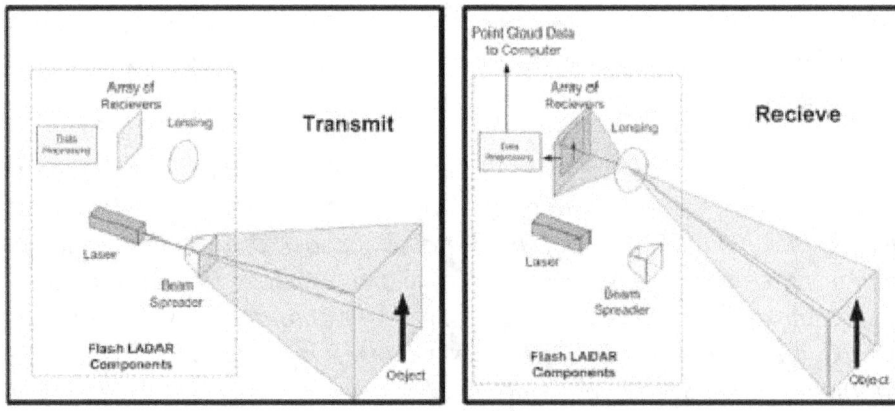

Figure 34. Illustration of ASC Flash LIDAR camera operation. (by permission from Applied Research Laboratory, Penn State University).

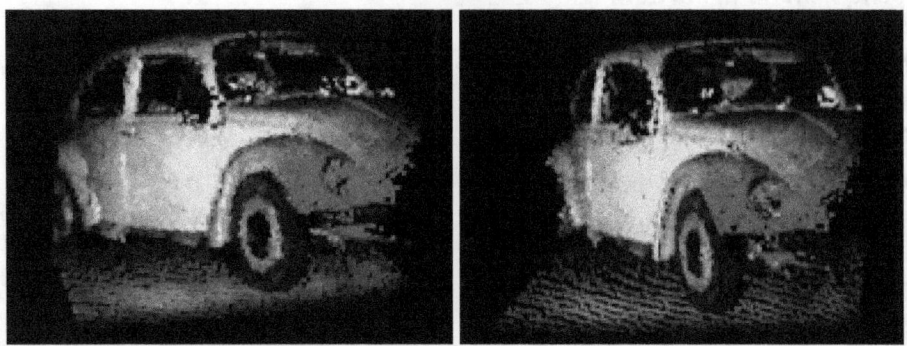

Figure 35. Two raw data image orientations of a single point cloud frame from an ASC 3D FLVC; 128 x 128 pixel resolution, color-encoded for range and intensity.

48

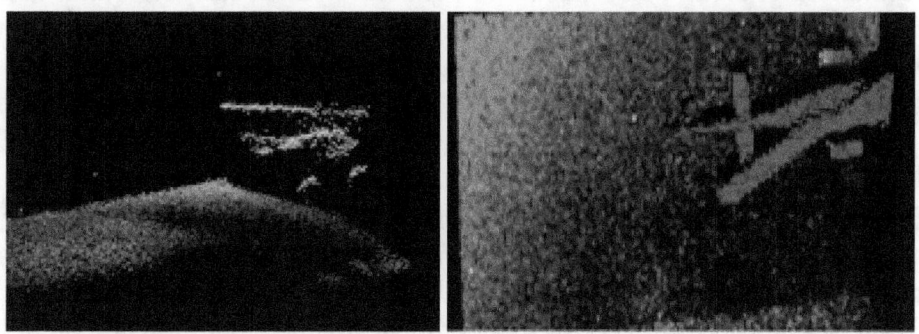

Figure 36. Two 3D raw data orientations of a helicopter using the same data; captured 396.24 m from above; color-encoded for range and intensity.

C. Scanning type real-time 3D imagers

A few developers continue to provide and improve real-time scanning type 3D imaging systems for mobility applications. They have been extensively tested in field operations and have demonstrated reliable and robust obstacle detection/avoidance and autonomous on and off road driving under difficult environmental conditions. Although they are more expensive than real-time FPA LADAR sensors, they have a much wider horizontal and vertical FOV which improves peripheral perception capabilities at great distances – especially when operating at higher vehicle speeds. All the current available sensors use the laser pulse TOF range measurement approach because of their reliable long range measurements under bright outdoor ambient conditions. Another feature that makes the pulse laser approach attractive is the ability to extract the range profile (multiple-returns) for a single beam column. The sensors that utilize this feature can improve range measurement performance in the presence of obscurants such as rain drops, dust, fog and foliage. Some of these imaging systems are described below.

(1) Sick Inc.

The simplest sensor system is a single plane wide FOV scanner available from Sick. Sick has been successful in marketing this type of sensor for industrial AGV and automated material handling systems for many years. Their LMS 100, LMS 200, and LMS 400 series offer different capabilities for various industrial indoor applications: Measurement range from short range (3 m) to long range (80 m) operation, a FOV from 70° to 270° operation, angular resolution from 0.25°, 0.5° and 1.0° per measurement step, scanning frequency from 25 Hz to as high as 500 Hz, range measurement uncertainty of ± 4 mm to ± 30 mm. Typical range resolution for all of the LMS series sensors is approximately 1 mm.

For outdoor applications, Sick has introduced a new multi-planar, long range, laser based 3D imaging system. It is based on the multi-planar scanning sensor concept developed by the IBEO Co. in Hamburg, Germany for automotive applications. The LD-MRS features an operating range of 0.5 m to 250 m, an 85° FOV when using four measurement planes and an 110° FOV when using only two measurement planes, selectable angular resolutions of 0.125°, 0.25°, and 0.5° per measurement step, selectable scanning frequency of 12.5 Hz to as high as 50 Hz, range resolution is 40 mm over the range of operation, and multiple pulse technology for improved outdoor capabilities including obscurant penetration.

(2) IBEO Automobile Sensor GmbH

IBEO, in Hamburg, Germany, offers the LUX Laserscanner for automotive safety system applications. This sensor uses a four or eight layer/planar scanning approach where the reflected returns are processed in parallel as the mirror scans the scene horizontally. These sensors perform very well under most weather condition and on hilly roads as long as the vehicle doesn't experience large vertical excursions such as can be expected in off road terrain. Typical range of operation for this sensor is between 50 m to 60 m; however, objects perpendicular to the road surface have been detected at 200 m in demonstrations. The unique feature of the sensor is the multi-target capability (detection of 3 return reflections from one laser pulse. This feature provides for reliable object detection even in the presence of obscurants (such as rain drops). Figure 37 shows how the multi-target feature operates.

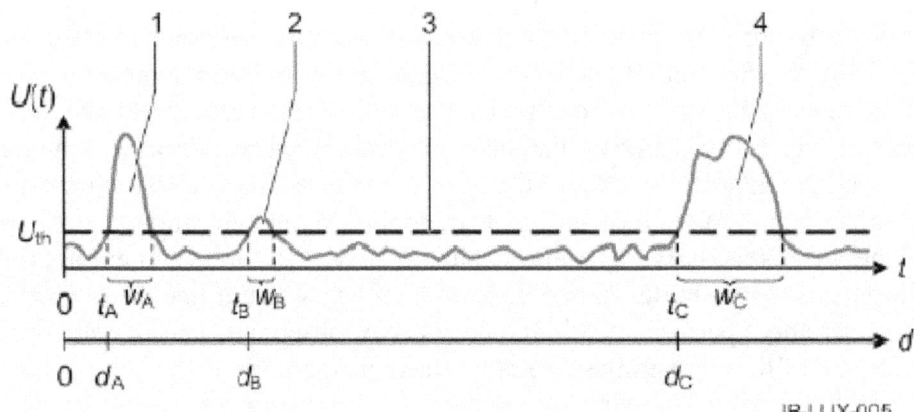

Figure 37. LUX Laserscanner multi-target return feature.

Return pulse (1) is the return pulse from the window pane. It is characterized by a high voltage signal, but over a very short period of time. Return pulse (2) is a reflection from a rain drop. It usually characterized by a low voltage pulse and lasts for a very short time. Voltage (3) is the threshold voltage setting and return pulse (4) is the last return from a solid object on the road. It is characterized as a high voltage return signal and lasts for a much longer time interval.

Figure 38 shows how the four layer/plane scanner is configured to provide a 3.2° vertical FOV capability. This feature allows the sensor to detect objects better on hilly roads and if the vehicle is experiencing heavy acceleration or braking.

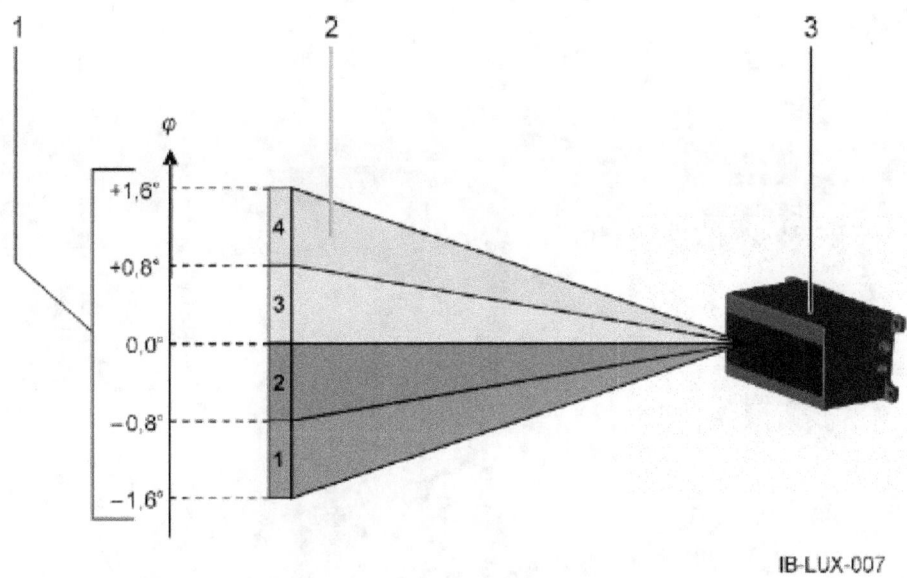

Figure 38. LUX Laserscanner four layer scanning configuration.

The LUX Laserscanner can be operated at three different scanning frequencies (12.5 Hz, 25 Hz, or 50 Hz). The horizontal angular resolution is based on the scanning frequency. The horizontal FOV (working area) for the four layer/plane configuration is 85° and 100° if only two layers are used. Multiple sensors can be used to expand the working FOV. Additional information is available at the IBEO website (http://www.ibeo-as.com).

(3) High definition laser based 3D imaging systems for autonomous on and off road driving. A couple of laser based 3D imaging systems have been designed specifically for autonomous on and off road driving. These sensors provide a very dense (high spatial resolution) point cloud and can achieve measurement rates of a million samples per second or better. This is necessary because the sensors need to cover a very large horizontal FOV (up to 360°) and vertical FOV (20° to 30°). This high density spatial resolution feature can be accomplished using a large number of laser/emitter pairs, where each pair establishes a layer/plane having high density (closely spaced) range measurements.

A commercial version of this type of sensor, manufactured by Velodyne, was utilized by many of the teams who competed in the 2007 DARPA Urban Challenge (http://www.DARPA.mil/grandchallenge/index.asp). The HDL-64E laser scanner was used on five of the six finishing teams, including the winning and second place team vehicles. Figure 39 shows the mechanical configuration of the sensor and Table 4 lists the sensor specifications (http://www.velodynelidar.com). In addition to autonomous vehicle navigation, the sensor can be used wherever high definition mobile and dynamic 3D imaging is needed, such as: automotive safety systems, mobile surveying (road infrastructure, rail systems, mining), mobile mapping, and security applications. Lockheed Martin has selected the HDL-64E as the main perception sensor for navigation and collision avoidance on the Squad Mission Support System (SMSS) robotic vehicles being built for Army future robotic weapon system needs.

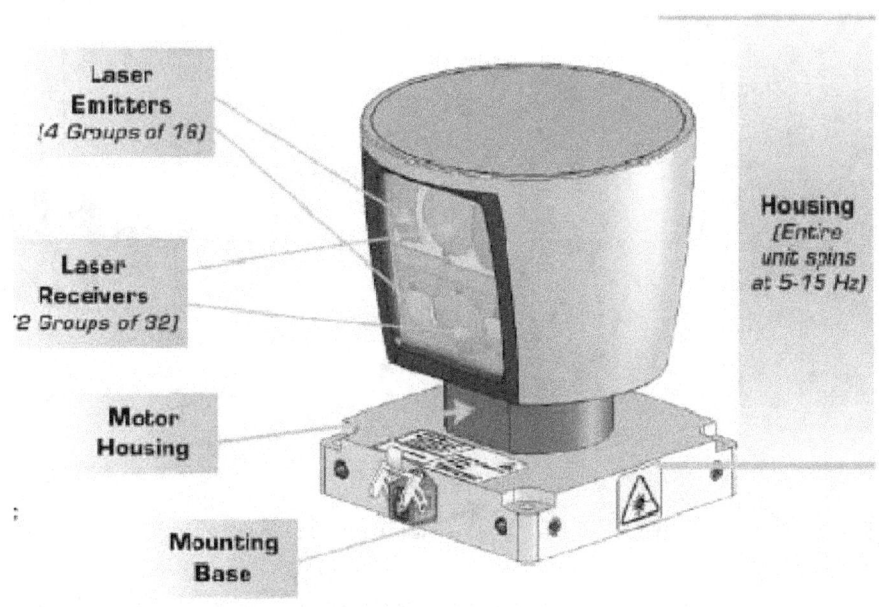

Figure 39. Mechanical configuration of the Velodyne HDL-64E S2 high definition 3D imaging system.

Table 4. Specifications of the Velodyne HDL-64E 3D imaging system

Sensor:	64 lasers / detectors
	360° FOV (azimuth)
	0.09° angular resolution (azimuth)
	26.8° vertical FOV (elevation) ±2° up to -24.8° down with 64 equally spaced angular subdivisions (approximately 0.4°)
	≤ 2 cm distance accuracy
	5 Hz to 15 Hz FOV update (user selectable)
	50 m range for pavement (≈0.10 reflectivity)
	120 m range for cars and foliage (≈0.8 reflectivity)
	$> 1.333 \times 10^6$ points per second
	< 0.05 ms latency
Laser:	Class 1M – eye safe
	4 x 16 laser block assemblies
	905 nm wavelength
	5 ns pulse
	Adaptive power system for minimizing saturations and blinding
Mechanical:	12 V input (16 V max) @ 4 A
	< 13.2 kg (29 lbs)
	254 mm (10 in) tall cylinder of 203 mm (8 in) OD diameter

	300 RPM to 900 RPM spin rate (user selectable)
Output:	100 Mbps UDP Ethernet packets

In August of 2010, Velodyne introduced the HDL-32E LIDAR sensor. It meets the demand for a smaller, lighter and less expensive high definition laser scanner for autonomous vehicle navigation and mobile mapping purposes. The following description of the new sensor was taken from a Velodyne press release:

> *"The HDL-32E extends the core technology developed for the revolutionary HDL-64E introduced in 2007. The HDL-32E measures just 5.9 inches [150 mm] high by 3.4 inches [86 mm] wide, weighs less than three pounds and is designed to meet stringent military and automotive environmental specifications. It features up to 32 lasers aligned over a 40 Vertical Field of View (from +10° to -30°), and generates 800,000 distance points per second. The HDL-32E rotates 360 degrees and provides measurement and intensity information over a range of five centimeters to 100 meters, with a typical accuracy of better than +/ 2 cm. The result is a rich, high definition 3D point cloud that provides autonomous vehicles and mobile mapping applications orders of magnitudes more useful environmental data than conventional LiDAR sensors."*

A military version of a high definition 3D imaging system , built by General Dynamics Robotic Systems (GDRS), has been used extensively in Future Combat Systems UGV test scenarios. It has demonstrated reliable and robust performance for autonomous driving and obstacle detection/avoidance as part of the ARL's Robotic Collaborative Technology Alliance program (http://www.arl.army.mil/www/default.cfm?page=392). Because the sensor technology is considered to be critical technology by the military, GDRS was not able to provide updated information on the specifications and performance of their latest generation models and on current research and development activities. The latest information available came from the 2003 Army CTA conference. At that time, a Gen IIIb LADAR was in development which had the following features and specifications:

- Eye safe laser wavelength of 1550 nm
- Maximum range of 100 m with 20 % target reflectance
- Maximum range of 50 m with 4 % target reflectance
- Large sensor optics 40 mm for improved sensor sensitivity and resolution
- 180 x 64 or 360 x 128 pixels per frame for a 120° x 30° FOV
- Range resolution of 2 cm
- Digital sampling of multiple pulse returns history
- 30 Hz frame rate for 180 x6 4pixels per frame, 15 Hz for 360 x 128 pixels per frame

D. MEMS based scanning-laser 3D imagers

An aspiration for a laser based vision system for mobility and machine automation was given in the 2004 NIST report [1]. The ideal attributes of such a system were:

- Illumination source   Eye safe laser (1500 nm)
- Field of View $90°$ x $90°$; $40°$ x $90°$; $9°$ x $9°$
- Range resolution   1 mm @ 15 m; 3 mm @ 5 m to 100 m
- Angular resolution $< 0.03°$
- Frame rate   > 10 Hz
- Size   "Coffee Cup"
- Cost   < $1000 US dollars

LightTime, LLC has taken on the challenge, by conceptualizing and developing a MEMS-based 3D imaging system (as a next generation LADAR) to achieve the goal of building a high accuracy 3D imaging system which is smaller, faster, cheaper, and amenable to high volume production [1, 59]. The initial objective is to build and demonstrate a fast frame rate camera for machine control, automatic field metrology and military applications having an operating range of less than 100 m. Working with military and automotive technical personnel, LightTime generated a set of performance specifications for such a MEMS- based system. The resulting high-level specifications for the MEMS sensor and mirror are listed in Table 5.

Table 5. High-level Specifications of MEMS Sensor and Mirror

Parameter	Spec Value	Notes
Scanner		
Maximum Frame Rate (fps at 100 m)	15 fps	or greater with tradeoffs in # pixels per frame
FDA Laser Classification (Class)	Class 1 (Eye Safe) - Optional	
Beam Divergence (diameter at specified distance)	1 mrad	(e.g., 50 mm at 50m)
Maximum Range (m) (@object reflectivity %)	100 m (@100%) 30 m (@10%)	
Scan FOV (vertical) Optical	$24°$ to $48°$	depends on MEMS mirror characteristics
Scan FOV (horizontal) Optical	$32°$ to $64°$	depends on MEMS mirror characteristics
MEMS Mirror	**Spec Value**	**Notes**
Mirror mechanical rotation angle	$\geq 8°$ x $6°$	minimum mechanical <u>full</u> angle (fast x slow axis)
Raster fast axis operating frequency	2 kHz	approximate value
Raster slow axis operating frequency (sets frame rate)	10 Hz	approximate value
Mirror scan motion	2D linear raster	synchronized

After investigating the market for the commercial availability of 1D and 2D MEMS scanning mirrors, LightTime found that none could meet their performance requirements of (1) mirror size, (2) angular range, (3) operating frequencies, and (4) highly linear activation. LightTime decided to develop a "second generation" 2D MEMS mirror system for the MEMScan prototype LADAR that they were developing. Figure 40 illustrates how a 2D MEMS mirror scanner would operate.

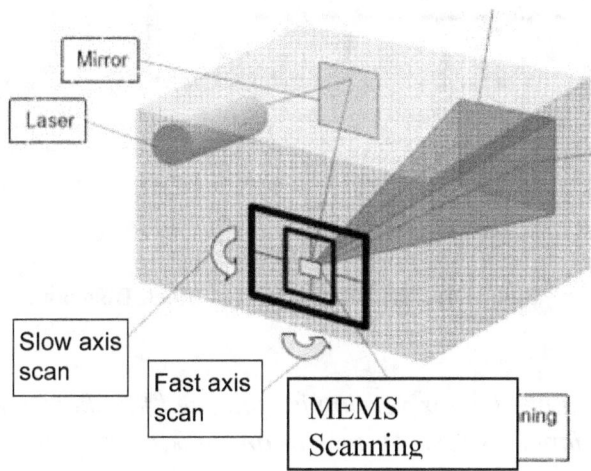

Figure 40. Conceptual illustration of a 2D MEMS mirror scanner in operation (courtesy of LightTime).

LightTime provided the following overview of their MEMScan program:

"LightTime recognized that it would first have to develop the appropriate "Second Generation" MEMS mirror system, and is doing so as an integral part of its MEMS scanning LADAR program. LightTime is presently developing a <u>prototype</u> of its new LADAR product, MEMScan^{TM}, a scanning-laser real-time 3D image sensor. MEMScan will be a compact, lightweight image-data capturing front-end component that interfaces with customers' image processing and control systems. The latter will post-process MEMScan's output data for the purposes of image display, analysis, and/or autonomous system control. LightTime intends to make MEMScan available as an OEM (original equipment manufacturer) component."

Figure 41 is system block diagram of the MEMScan top-level design. The dashed line represents the path of a single (emitted and reflected) laser pulse.

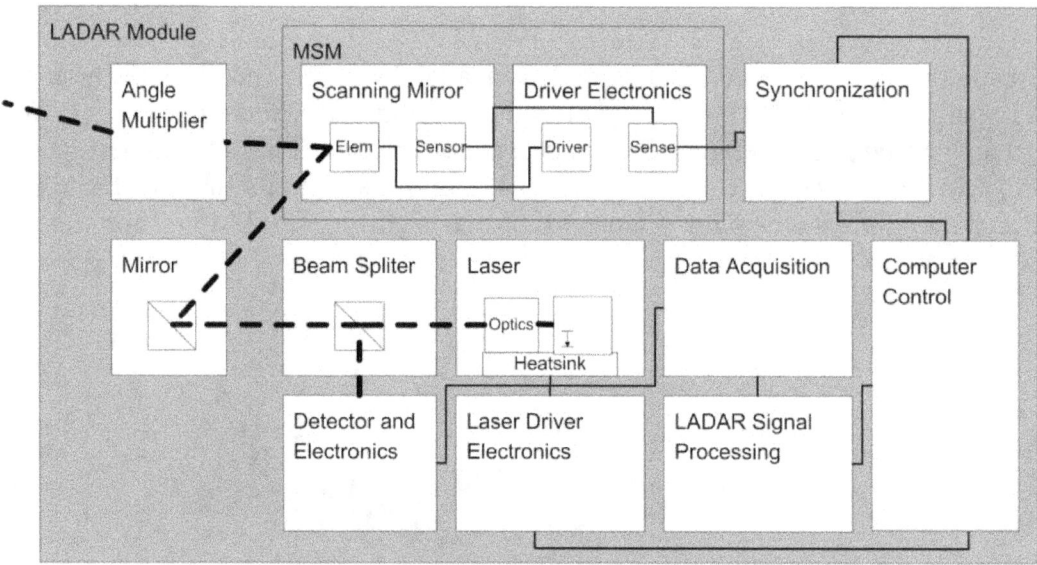

Figure 41. MEMScan System Block Diagram.

"The initial MEMScan production version is optimized for short ranges (1 m to 100 m); however, its design embodies several innovations making it scalable to longer-range applications. MEMScan's key design element is the use of a proprietary 2D MEMS scanning mirror where, due to the mirror's design in combination with special optics, the mirror's native scanning range is increased from several degrees to producing at least a 32° (horizontal) optical field of view; with potential of up to 64°.

Currently, the mirror is in the final stages of hardware test, targeted for sampling in Q1 2011. A complete LADAR prototype is targeted for sampling in Q4 2011."

Features of the MEMS scanning approach are:

- Real time 3D image data capture (data generated for each pixel in the field-of-view includes: range, intensity, slow and fast axes angles)
- Miniature system; without batteries/converter: < 400 cm^3 and < 0.5 kg
- Raster scan pattern for efficient operation and simple standardized upstream image-data processing
- IR scanner design that is low footprint
- Eye safe operation
- Possible cost of less than $100 in large volume production

In a paper published in the 2006 SPIE Defense & Security proceedings, LightTime made the following statement concerning potential applications for the next generation MEMS LADAR technology [60]:

"This next-generation of MEMS LADAR devices is not anticipated to replace the relatively mature current technologies initially but rather be used for new applications where current LADAR devices are not appropriate because of size, weight, speed, or their inability to

withstand rugged conditions. Such real-time applications include unmanned aerial vehicles (UAV), helmet mounted LADAR for the individual soldier[61], and many more as described in Table 6."

Table 6. Potential New Applications for MEMS Scanning LADAR

Area of Application	Description/Advantages
Remote Surveillance (border/perimeter)	Low power consumption and small size (easier to hide/protect). Could be used to protect a border for significantly less than $1M/mile.
Reconnaissance	They could be dropped behind enemy lines for quick intelligence and monitoring.
Missile guidance	Real-time 3D imaging. Short-range or as an adjunct (limited to < 1 km).
3-D Mapping	Real-time 3D battlefield imaging, mission planning, virtual reality simulations, etc.
Target Acquisition/ID	x, y, z of target for soldier targeting and/or remote fire control/guidance.
Automotive	Obstacle identification & autonomous navigation under rugged conditions.
UAV	Perfect for aerial drones since it is small, lightweight, and has low power consumption.
Terrestrial Drones/Robots	Though size and weight are not as critical as with an UAV, they are still important as well as power consumption.
Medical	Laparoscopic surgery, robotic surgery, face/body 3D collection (for virtual manipulation or reconstruction prior to surgery), etc.
Manufacturing	Robotic manipulation and depth of field vision, 3D bar code reading, etc.
Civil Engineering	Real-time quality control (e.g., aggregate analysis for road beds).
Scene Archiving	Forensic, archeological, etc. (small & lightweight, can go anywhere).
Movie/Camera Industry	Concurrent range data incorporated into each pixel of the digital image for computer processing into a 3D movie (TV, video, etc. as processing becomes more mature).
Simulation/Gaming	Real-time, real-word input of 3D scenes and actions for simulations/game design.
Scientific	Application such as particle analysis and distribution, microscopy.
Other	Laser printing, etc.

The ARL has also reported on the development of a MEMS-based 3D imaging system for use on small military UGVs (PackBot size vehicles). This LADAR research program is a near-term effort to build a LADAR using COTS components and mount it on a small ground robot and thus support autonomous navigation research [40, 62]. An early prototype has been built and tested, and shows great promise that the MEMS approach will provide for a low cost, compact, low power 3D perception solution for small robot applications. The design approach is similar to LightTime, except, that it is entirely based on using COTS components. ARL uses a fiber laser which operates at a rate of 200K laser pulses per second and uses a commercially available 1.6 mm diameter MEMS mirror vs. the custom designed (4 x 4) mm MEMS mirror designed by

LightTime. The following is a description of the system architecture which was provided by ARL:

"The LADAR uses a pulsed laser to determine range to a pixel and a two-axis MEMS mirror to establish the angular direction to a pixel. The LADAR architecture is depicted in the block diagram of Figure 42; detailed descriptions of the LADAR are included in the attached references [62]. Referring to Figure 42, a trigger signal commands an erbium fiber laser to emit a short 2 ns to 3 ns pulse of light at a rate of 200 kHz that is collimated and then directed to the surface of a small MEMS mirror. Analog voltages from a high-voltage (HV) amplifier set the pointing direction of the mirror. Light reflected from the mirror then passes through a telescope that "amplifies" the scan angle of the MEMS mirror. Light backscattered from the target is collected on the large face of a tapered fiber bundle that effectively increases the diameter of the photo detector, thereby improving the signal-to-noise (S/N) by a factor of three. Photocurrent from the detector is fed into a monolithic 50-ohm microwave amplifier. This output is then split into a low and high gain channel. On the radio frequency (RF) interface board, the low and high gain channel outputs are each summed with the photocurrent from a detector illuminated by an undelayed sample (called T-zero) of the original transmitted light signal. The T-zero pulse of the transmitted signal may then be used as a reference to determine target range. The outputs of the two channels feed respective inputs on a two-channel 8-bit analog-to-digital convertor (ADC). The ADC samples at a rate of 1.5 giga-samples-per-second (GSPS) and a first-in, first-out (FIFO) register is commanded to start acquiring ADC data upon transmission of the laser pulse by signals from a field programmable gate array (FPGA). The FPGA stores the amplitude data as a function of time from the ADC, determines the range to the pixel, and formats the data for acquisition by a PC for display. The FPGA also controls the pointing direction of the MEMS mirror and directs the laser to emit a pulse."

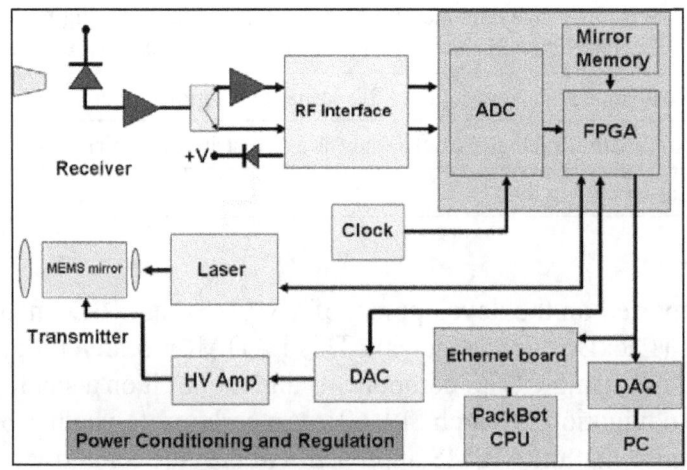

Figure 42. ARL MEMS LADAR system block diagram.

The LADAR was built to be mounted on the iRobot Explorer PackBot shown in Figure 43. The desired requirements specify that the LADAR fit within the PackBot sensor head [229 mm x 210 mm x 51 mm (9 in x 8.25 in x 2 in)]. However, since the sensor is an early demonstration prototype, only the transmitter and receiver parts are located in the sensor head. The control, data collection and signal processing hardware are located in the payload area. The initial desired performance requirements of: 6 Hz frame rate; a 60° x 30° FOV and an image size of 256 (h) x 128 (v) pixels; 20 m operating range; eye safe laser illumination; and at least a 40 cm range resolution was achieved with the current prototype.

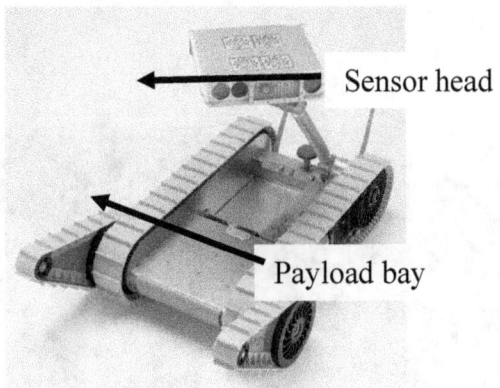

Sensor head

Payload bay

Figure 43. Explorer PackBot.

In April of 2010, ARL reported on 2009 achievements at the 2010 SPIE Defense & Security conference [40].

> *"This year we started redesign of the transmitter assembly so that the entire structure will fit within the PackBot sensor head. We replaced the commercial mirror driver board with a new small in-house design that now allows us to attach the top to the PackBot sensor head without a modification that increased its depth by 10 mm (0.4 in). We built and installed a new receiver that is smaller than the original and possesses three-fold higher signal-to-noise with roughly twice the bandwidth. Also we purchased and tested a new Erbium fiber laser that requires a fraction of the volume and power of the laser used in last year's effort. The most significant work focused on software development that integrates the LADAR system with the PackBot's onboard computers. This software formats the LADAR data for transmission over the PackBot's Ethernet and manages the flow of data necessary to drive the PackBot and transmit LADAR and video imagery over a wireless connection to a laptop computer for display. We conducted tests of the entire PackBot and LADAR system by driving it over a short course in the lab while displaying and recording video and LADAR data."*

In the publication, ARL provided a review of the LADAR architecture, briefly described the design and performance of the various subassemblies and software, summarized the test activities, and identified future development work.

A driving test of the sensor mounted on the PackBot was conducted at the ARL lab. Figure 44 is a photo snap-shot of the test bay. Figure 45 shows one LADAR frame from the run in grey scale, and Figure 46 shows the corresponding LADAR image with false-color range information. The test results revealed that the alignment between the right and left motion scans of the MEMS mirror were slightly out of place on vertical edges of objects in the scene. By adjusting the left and right motion scans relative to one another, they were able to considerably reduce the misalignment. In the next phase of the development, they expect to increase the S/N by about six-fold with the installation of a new receiver and a better laser to obtain even better range image quality.

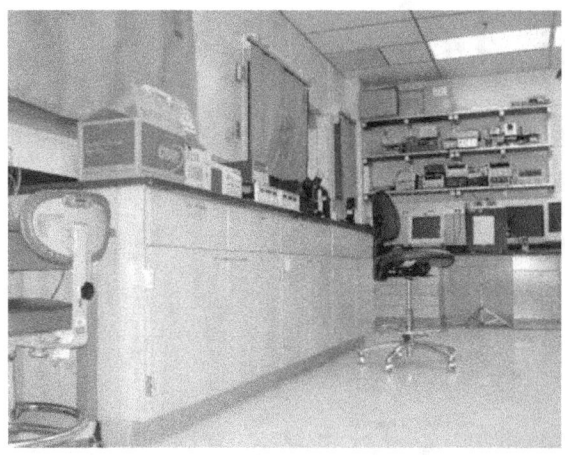

Figure 44. Photo snap-shot of lab test scene.   **Figure 45. Grey scale LADAR image.**

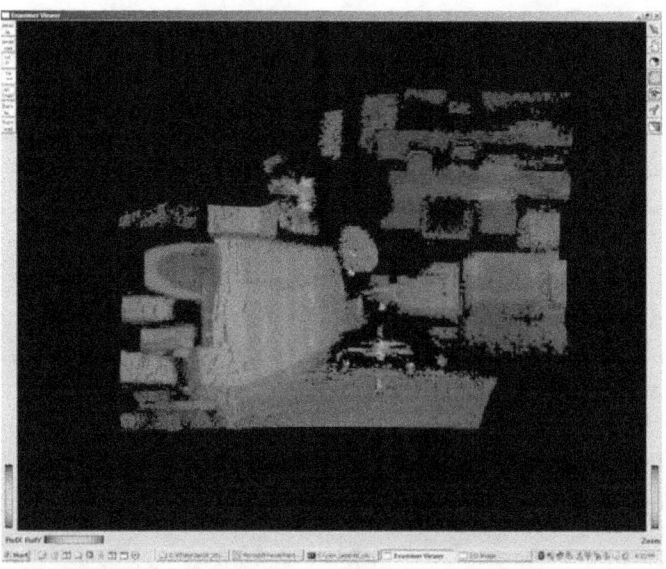

Figure 46. False color range LADAR image.

3.2.2.5 Advances in electron avalanche photodiode (e-APD) FPA technology

The Infrared Technologies Division of DRS Sensors and Targeting Systems has reported that the next generation of IR sensor systems will include active imaging capabilities [63]. Early development work on a new high gain HgCdTe electron avalanche photodiode was reported in 2003 and 2004. This work was funded by the DARPA MEDUSA program. Since then, progress has been made in implementing a 128 x 128 FPA for a gated – active/passive imaging system which is primarily designed for targeting and identification purposes at fairly long ranges.

In [63] DRS describes that:

> *"The Gated FLIR (forward looking IR) is an active imaging system that uses a laser pulse to illuminate the region of interest in a scene. The reflected radiation is detected by a camera that is gated to integrate signal only during the time period when the return pulse is expected. This gating improves signal to noise and removes scene clutter due to reflections off objects that are either in front of or behind the object of interest. Target ID is enhanced due to the high resolution and low FOV of the active imager."*

Typically, a WFOV (wide FOV) daylight or thermal imaging system is used as a search engine for targeting, before imaging is done with the high resolution gated FLIR. This requires precise alignment for accurate registration of the target information from the two systems.

DRS offers the following solution for a simplified targeting system:

> *"The solution is a combined sensor wherein the same FPA and optical system is used for both passive and active operating modes. This eliminates the need for bore-sighting two different cameras and greatly reduces the overall complexity and size of the system. The HgCdTe MWIR (Mid Wave Infrared) electron avalanche photodiode (e-APD) provides an elegant and high performance solution to the sensor in that it can be switched from active to passive modes by simply changing the bias on the detector. Obviously, the readout and the optics must also accommodate the two modes of operation."*

The following is a description of the detector architecture and performance of the HgCdTe APD provided by DRS:

> *"The HgCdTe avalanche photodiode (APD) detector design is based on the ... "high density vertically integrated photodiode" (HDVIP™) architecture developed by DRS. The HDVIP™ structure forms a front-side illuminated, cylindrical, n-on-p photodiode around a small via (circular electrical connection) in the HgCdTe, as shown in Figure 47. Details of the structure, operating theory, fabrication processes, and measured performance of these devices have been previously reported [64-66].*
>
> *The via provides electrical connection between the n-side of the photodiode and the input to the readout circuit. The HDVIP™ structure is currently employed at DRS for*

production of SWIR (Short Wave Infrared), MWIR, and LWIR (Long Wave Infrared) staring arrays. Important features of this design include: (1) interdiffused CdTe passivation of both surfaces for low 1/f noise; (2) thermal cycle reliability that is detector- and array-size independent; (3) low defects due to near-parallel diode junction orientation with respect to threading dislocations; and, (4) front side illumination for high quantum efficiency, high fill factor, and good MTF (modulation transfer function).

The optically active area is the total pixel area (pixel pitch squared) minus the area of the via and the area of the top side grid substrate corner contact, if used. The area of the via is typically 95 % to 98 % of the total area. Without a grid, the fill factor for a 40 μm pitch pixel with a 6 μm via is 98 %. A half grid corner contact for a 40 μm pitch single 6 μm via pixel results in an optical fill factor of 84 %.

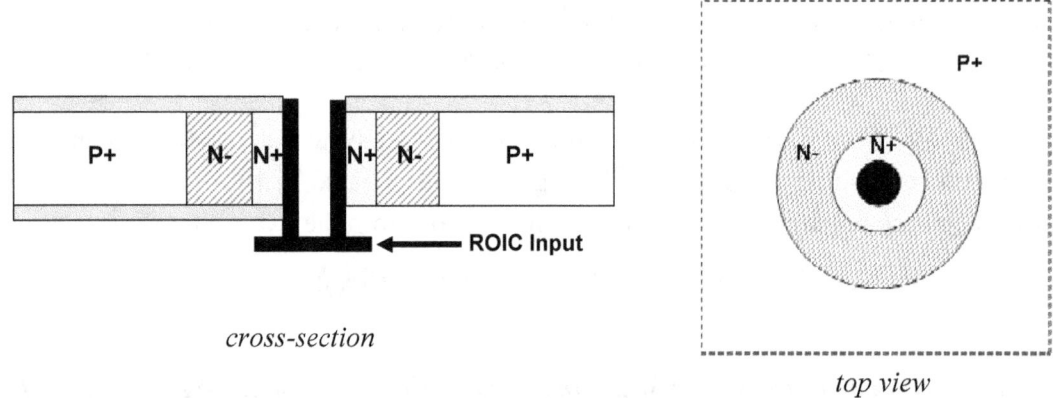

cross-section

top view

Figure 47. HDVIP^{TM} e-APD architecture.

The excess noise of an APD is important in determining ultimate performance. Because the gain process in an APD is random, there is an excess noise that is associated with the variance in the gain. The cylindrical geometry of the HDVIP™ diode strongly favors electron injection and multiplication over the corresponding vacancy (hole) processes. This inherent carrier selectivity, coupled with the unique band structure of the HgCdTe material, and the electron mobility and lifetime characteristics at cryogenic temperatures, result in extremely low (near unity) excess noise for MWIR devices. More importantly, the excess noise is independent of gain, remaining low even as gain is increased to very high levels. Figure 48 shows measured gain and excess noise factors for a representative MWIR device, confirming an excess noise factor of unity to gains as large as 1000."

The derivation of excess noise as a function of *k* was provided by McIntyre [67] where k is the theoretical hole-to-electron ionization coefficient constant in electron-photon interactions. A comparison of the measured excess noise factor to the theoretical coefficient is provided in [68].

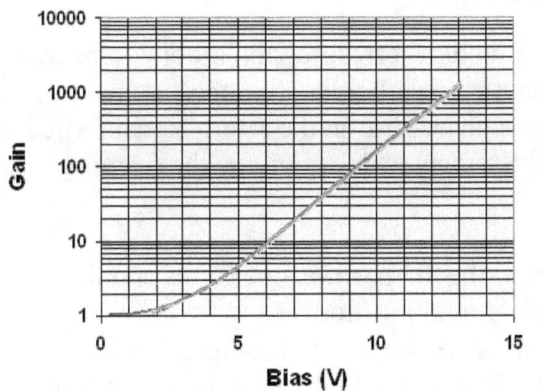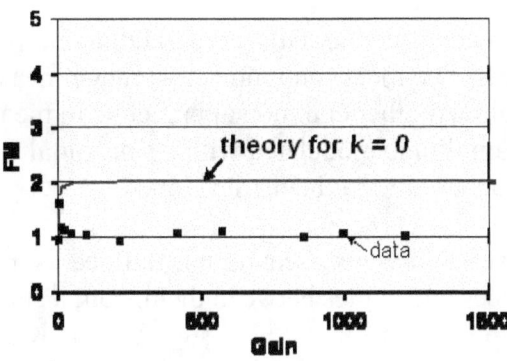

Figure 48. The gain (left) in MWIR HDVIP devices can exceed 1000 with less than 15 V of bias, and the excess noise factor remains near unity even at high gain.

Details of the FPA design, the ROIC design, the FPA performance, the demonstration system, and test results are provided in the 2007 SPIE paper on the gated FLIR FPA [63].

An example of the gated FLIR system performance is shown in Figure 49. An SUV was targeted at a distance of approximately 0.5 km from the sensor. The gated image is quite sharp with a lot of detail so that it can be easily identified as an SUV. The gate width for the sensor was set to approximately 200 ns (or about 30.5 m [100 ft] in range).

Figure 49. An SUV at 0.5 km was used as the primary target for the initial outdoor test. A fan resolution target was also included to provide a quantitative assessment of image quality.

The demonstration test confirmed the capabilities of the actively gated FPA device to support both passive MWIR and active SWIR operating modes and demonstrates the potential of developing advanced sensor systems for future gated imagers and possibly for high resolution laser based 3D imaging.

3.2.2.6 Terrestrial and mobile terrestrial 3D imaging

The greatest advances in active 3D imaging products include terrestrial and mobile terrestrial systems. The most common active range measurement approach utilized in these sensors is time of flight. Another common approach is to measure the phase shift in the return signals which has been amplitude modulated with a sinusoidal signal. Systems which combine the features of both approaches are also being developed.

Terrestrial surveying/scanning is defined as measuring the 3D positions of points and the distances and angles between them from a stationary ground position (http://en.wikipedia.org/wiki/surveying). Mobile terrestrial surveying is similar but the measurement is from a mobile reference point established with a high-accuracy navigation position sensor that uses both GPS (Global Positioning System) and INS (Inertial Navigation System) information. Terrestrial sensors currently are the primary scanning sensors selected for applications in industrial reverse engineering, product inspection and modeling, industrial productions facility modeling and mapping, terrain mapping, and in heritage preservation. Mobile terrestrial sensors are being used in transportation related applications such as: surveying, highway design, corridor development, highway safety.

The hardware technical survey information, on terrestrial type scanning sensors for this study, is incomplete. Only four of the main thirteen commercial developers offered to provide information for this study and survey. Refer to Appendix B: Hardware technical survey for the provided information. However, as evident in the previous sections of the report, there was a very positive response from newer companies introducing new active 3D imaging products or research groups developing new measurement concepts. The reason for this lack of interest by the more established companies may be because they believe that the GIM International and Point of Beginning surveys are sufficient for their marketing needs. The reader should go to the corresponding websites (http://www.gim-international.com, http://www.pobonline.com) to obtain details of sensor features and specifications available from the following producers: 3rdTech, Basis Software, Callidus, FARO, Leica, Maptek I-Site, Measurement Devices Ltd., Optech, Riegl, Spatial Integrated Systems Inc., Topcon, Trimble, and Z+F.

Ongoing research information at these companies was also very difficult to obtain. Three of the four companies who participated in the study, said that no major changes were expected in the current products, however, the companies are continuing to enhance sensor performance by lowering measurement uncertainty, improving dimensional accuracy, increasing scanning rates, improving sensor functionality and speed of operation, and lowering cost. It appears that making advances in software processing of range image data is a much more important development effort for terrestrial 3D imaging developers. This may be driven by the needs of service companies for better and more automated processing software.

An overview of 3D laser scanners used for high resolution mapping of all types of facilities, structures, utilities and terrain is presented in Chapter 10 of a manual prepared in 2007 by the US Army Corps of Engineers titled: "Engineering and Design: Control and Topographic Surveying" [3]. The following is a summary of the high points presented in the report:

- Terrestrial laser scanners can be used to scan objects at high density – over a designated FOV – at speeds upwards of 500K samples per second – having pixel dimensions smaller than 5 mm.
- Although 3D range accuracies as low as a millimeter are claimed by some developers (with redundant measurements), 5 mm to 10 mm uncertainties are obtained in practice.
- A full FOV scan of a site or structure can be performed in 5 to 15 minutes, but may require several setups to provide full detail coverage.
- The scanners are very useful for mapping electrical generating facilities, dams, archeological sites, traffic areas, and imaging hard to reach locations.

The report also covers the following important topics:

1. Product manufacturers: Terrestrial laser scanners came on the market in the late 1990s. As of 2005, about 12 manufacturers were listed in trade publications.
2. Cost: The cost of a complete laser scanning system (including modeling software and training) can run between $150K and $200K. Daily operating costs of $2,000 to $5,000 or more (including processing) are not uncommon.
3. Accuracy: Relative accuracies are very good (5 mm or better at close ranges). Absolute accuracies depend on the accuracy of the site reference network and how accurately the sensor is aligned to the network. Absolute accuracies can be kept within 1 cm to 2 cm over a small project/structure site.
4. Density of scanned points: 3D imaging systems can be set to any desired scan density, e.g., 5 mm to 1 m at a nominal measurement distance. The higher the density setting, the longer the data collection, editing, and processing time.
5. FOV: Typically scanners can be set to scan a full 360° horizontal field or to a smaller FOV setting. Vertical settings are typically less than 310°.
6. Range: Some scanners are designed for close range scanning of 200 m or less while other claim ranges of 1000 m or more. The longer the range, the larger the footprint and less accurate the resulting measurement. Most detailed scans of facilities and buildings are obtained at ranges of 150 m or less. Eye safety must also be taken into consideration – A longer range sensor may require a higher power laser (Class 3). This may not be acceptable for surveying in populated areas.
7. Beam footprint size: The footprint size varies with the distance of the object from the scanner. At 30 m or less, a 5 mm footprint can be observed.

Transportation agencies in the U.S. and worldwide have increasingly used 3D imaging technology, often in multi-sensor configurations that include GPS, photography, and video for various applications. The LIDAR Focus Group of the Wisconsin Department of Transportation is assessing the most promising uses of 3D imaging for transportation. Their recent report on "Lidar Applications for Transportation Agencies", released in February of 2010, represents an initial scoping of this topic [69]. This report provides selected resource information on: Principles of 3DI, Types of 3DI, 3DI applications, technical issues, and other resources for additional information.

The report provides sample citations for each application area that includes information on practicality of use, sensor performance analysis, and analysis of cost associated with actual case studies. The report also states that there is significant on-going research in the areas of: 3DI data collection and analysis, 3DI error and accuracy, and on integration of 3DI and photogrammetry.

3.2.2.7 Obscurant penetration and collision avoidance for helicopters

Helicopter brownout conditions can occur from sand, dust or debris being pushed upwards by the rotor downward thrust during take-offs and landings. It can cause the pilot to lose visibility of the ground reference terrain. Figure 50 provides an example of such a condition. It was obtained from an article by Gareth Jennings for IHS Jane's (www.janes.com) titled " Down in the dirt" for the military [70]. The following statements were highlighted in the report:

- The number of brownout-related accidents involving rotary-wing aircraft has more than doubled since US-led operations began in Afghanistan and Iraq.
- Finding a solution to the problem has become a priority for the military and industry.

Figure 50. Example of brownout conditions for rotary-wing aircraft (from IHS Jane's report).

Two companies addressing this problem provided the following write-ups and information on their concepts for penetrating obscurants in helicopter visibility:

1) The **Neptec Design Group** prepared a synopsis on their Obscurant Penetrating Autosynchronous LIDAR for this publication:

Overview

66

The Obscurant Penetrating Autosynchronous Lidar (OPAL) has been developed as a variant of Neptec's TriDAR laser scanning system used for space vehicle rendezvous and docking. OPAL is designed to penetrate obscurants such as dust, snow, fog to detect objects engulfed in a cloud of obscurants.

OPAL has been in development since 2007 and has been tested at various military facilities in view of providing a laser-based vision system to assist helicopter pilots operating in dusty areas. A pre-production TRL-7 model was completed in July 2009, and subsequently flight tested in January 2010.

The OPAL is a general purpose sensor designed for many applications where a high resolution 3D sensor is required to operate in all weather conditions where visibility can be reduced due to dust, snow, and fog.

Background
The development of OPAL started in 2007 as a lab prototype system. The prototype was initially tested for its obscurant penetration capabilities at DRDC (Defense Research and Development Canada) - Valcartier laser propagation facility, which is equipped with an instrumented dust chamber. Following that, it was also tested at Porton Down (Porton, England) for UK Ministry of Defense and it was tested in March 2009 at the Yuma Proving Ground (Yuma, Arizona) facility at the invitation of the US Air Force Research Lab. The formal tests were very successful and indicated a penetration capability in dust concentration of about 3 g/m^3, which is in the upper range of typical dust concentration produced by large helicopters attempting to land on dusty/sandy ground.

Neptec also received a contract from CAE in early 2009 to build a flight-worthy OPAL system which was completed in July 2009. The instrument is similar to the lab prototype but provides a higher sensitivity detector and a faster data acquisition rate at 40 kHz. The system was tested successfully in January 2010 aboard a Bell 412 helicopter at Yuma Proving Ground.

Current State of Technology Development
The dust penetration capability is achieved through the use of a proprietary optical design that minimizes detector saturation and through the exploitation of the trailing edge of the light return pulses, allowing the system to detect obstacles engulfed in a cloud of obscurants. Please refer to the following references to get a better understanding of the approach used in the OPAL design [71, 72]. In addition to the hardware, a filter has been developed and implemented in software to remove a large proportion of the laser pulse returns caused by atmospheric particulates. The net effect is to reduce the noise in the 3D imagery and detect the presence of objects and terrain in the obscured environment.

In terms of other obscurant penetrating technologies, radar has traditionally been the sensor of choice. However, radar suffers from a poor spatial resolution capability. In contrast, OPAL provides a spot size of 3 cm at 100 m range, offering the capability to detect small objects that can pose a threat to a landing aircraft, such as wires, rods, fence posts, and tree stumps.

Figure 51 illustrates an example of dust penetration using OPAL in an environment where dust has been generated by a UH-1 helicopter. Figure 52 shows an example of the application of the dust filter, where the light reflected by dust particles suspended in the air is removed from the raw data.

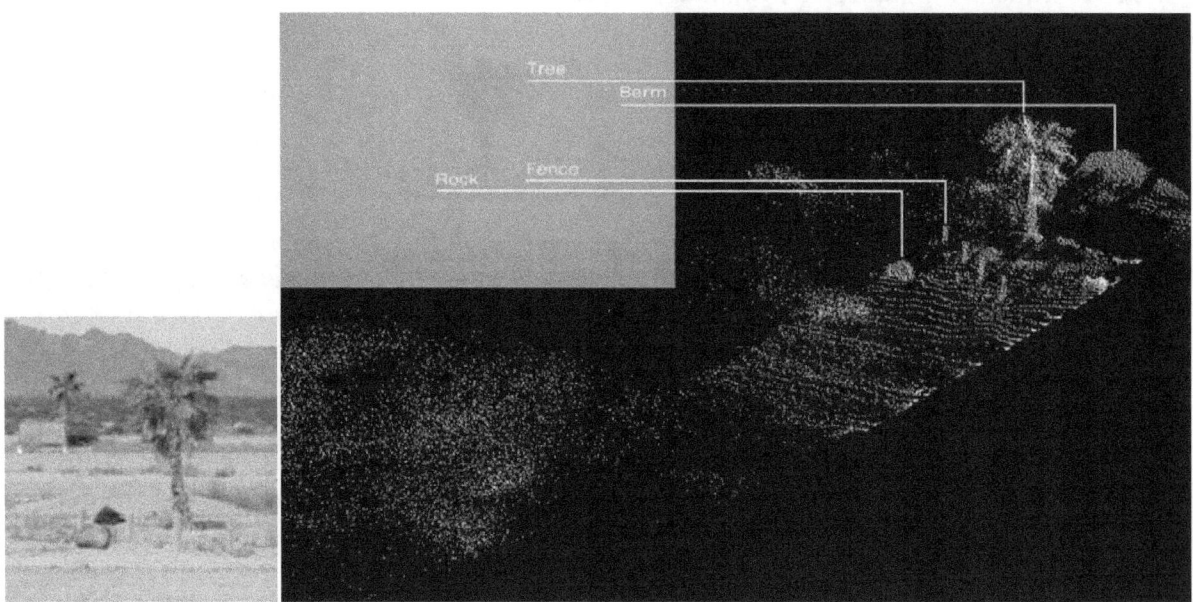

Figure 51. Example of dust penetration. Left picture shows the environment in clear conditions. Right picture show the raw OPAL data with a top left insert showing the same environment obscured with dust. Main objects are referenced between the picture and the LIDAR data.

Figure 52. Example of the application of the dust filter to raw data. Top left shows the scene prior to dust generation. Top right shows the scene after dust generations. The lower left shows the raw OPAL 3D data with false coloring: red corresponds to the ground, green to the targets and white corresponds to the noise caused by the light returns due to dust particles. The picture on the lower right shows the same raw data after the OPAL dust filter algorithm has been applied.

Performance characteristics for the OPAL are given in Table 7.

Table 7. TRL-7 OPAL Prototype Performance Data

Parameter	Value	Notes
Laser Wavelength	1540 nm	
Laser Beam Size	10 mm	
Laser Beam Divergence	0.3 mrad	
Field of View	30° x 30°	(H x V) The field-of-view can be slewed through a vertical range of 60°. The field-of-view size and aim is configurable through the scan command.
Angular Resolution	0.001°	
Data Acquisition Rate	up to 32,000 points per second	
Target Detection – Clear Air	2000 m for a 80 % reflective surface 550 m for a 6 mm (0.25 in) metal wire	
Target Detection – Obscurant	15m for a 80 % reflective surface in 5 g/m^3 uniform 10 µm diameter dust cloud.	

	50 m for a 80 % reflective surface in 1 g/m³ uniform 10 μm diameter dust cloud.	

2) **EADS Deutschland GmbH** – Defense Cassidian (EADS) of Germany has over twenty years of experience in developing and using LADAR technology. During the German military unmanned ground vehicles program, PRIMUS (Program of Intelligent Mobile Unmanned Systems, 1995 – 2004), they used a LADAR sensor to provide range information about the environment and were able to successfully detect and avoid obstacles in the path of the ground vehicle. Based on this experience, Cassidian developed an enhanced LADAR for helicopter collision detection/avoidance applications. The following description of helicopter collision hazards came from a paper presented at the 2008 SPIE Europe Security & Defense symposium in Cardiff, UK [73].

> *"Power lines and other wires pose serious danger to helicopters, especially during low-level flight. The problem with such obstacles is that they are difficult to see until the helicopter is close to them, making them hard to avoid. Even large pylons, towers and tall trees are tricky to spot and avoid if they do not stand out clearly from their backgrounds."*

To overcome these difficult problems, EADS Deutschland GmbH - Cassidian Electronics of Germany developed the HELLAS-Warning System (HELLAS-W) which was introduced to the market in 2000. The system uses a pulse modulation TOF scanning LADAR to detect obstacles in the flight path of the helicopter and provides a visual and audible warning signal to the pilot. The system is currently being used by the German Federal Police & Emergency Medical Services and is starting to be deployed on military helicopters.

Figure 53 shows a photo of the HELLAS Laser Radar scan head and a diagram of the operating principle.

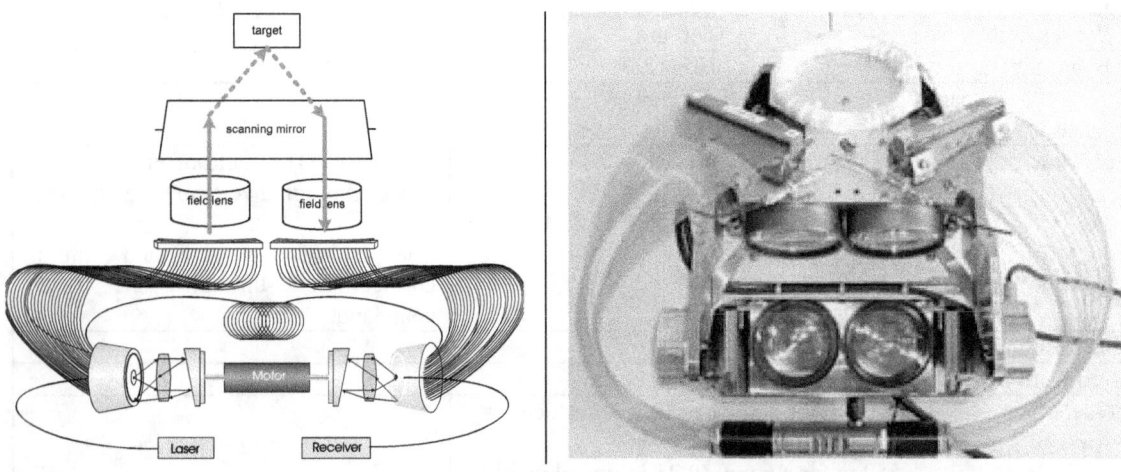

Figure 53. Photo of HELLAS Laser Radar scan head and diagram of operating principle.

The next generation HELLAS-Awareness System (HELLAS-A) which is shown in Figure 54, is targeted for use on military helicopters. It is currently undergoing performance qualification tests. The system uses the same LADAR architectural concept as the HELLAS-W, but has improved performance specifications and is an integral part of the aircraft avionics. HELLAS-A overlays the detected and identified hazardous obstacles over other sensor information, such as FLIR, and database information, such as digital maps. This is illustrated in Figure 55.

Figure 54. HELLAS-A.

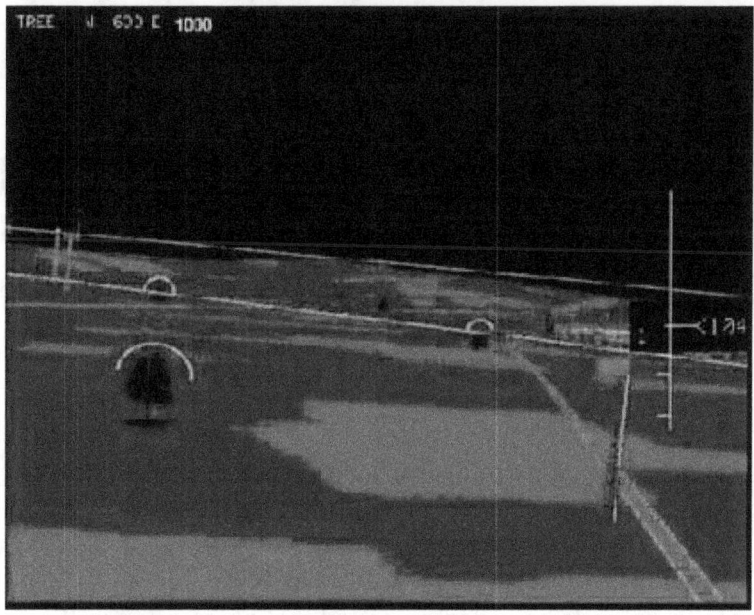

Figure 55. Obstacle symbology overlaid over FLIR video.

The HELLAS-A sensor is designed to detect a 5 mm wire at a distance of more than 700 m at a visibility of 12 km with a detection probability of 99.5 percent per second. This system can be fully integrated in the helicopter avionics system. Its results can be displayed on the helicopter's multifunctional display (MFD) units or on a pilot's head-mounted sight and displays (HMS/D).

HELLAS-A plans to address another common challenge. When helicopters land on sandy or dusty ground they experience "brownout" where their visibility is blocked by the disturbed sand, dust (Figure 56), or snow. Such situations are very disorienting for the pilots and can lead to accidents as reference points are obscured. HELLAS-A tackles this with brownout recovery, a three-dimensional see-and-remember system. It does this using its situation-awareness suite that has a range of different detectors, including radar and LADAR. The scene that is previously detected is kept in memory and is displayed during a brownout situation.

Figure 56. German CH53 in a brownout. Source: Heeresflieger Laupheim.

The detected image data of the landing area (collected by the HELLAS system) prior to brownout, is displayed to the pilot along with supplementary data from other sensors (such as millimeter-wave radar sensors) which can penetrate obscurants and detect objects and movement on the ground. This fused information provides the pilot with "enhanced synthetic vision", allowing the pilot to fly safely in zero (human eye) visibility, low light level, and in bad weather conditions [74, 75]. Initial tests have demonstrated that objects as small as 0.4 m in size can be detected in the 3D data. Figure 57 shows an example of the HELLAS see-and-remember vision display that is provided to the pilot.

Figure 57. HELLAS See-and-Remember Vision.

The HELLAS real-time 3D imaging technology has also been used to address the growing needs for safety and guidance of unmanned ground vehicles. Some of the robotic support functions that have been tested by Cassidian Electronics include:

- the detection of terrain / driving obstacles
- road detection and following
- building of a local and global obstacle situation map
- detection of negative obstacle (ditches/depressions)
- vehicle / object following
- detection of explosive trip wires in the path of the vehicle

3.3 Software

3.3.1 Background

The primary advantage of 3DI systems is the ability to acquire millions of 3D points in a short period of time and the amount of detail that can be extracted from the 3D data. However, good post-processing software is required to fully realize this advantage. Such software has to be able to read and display large datasets quickly. At a minimum, it has to provide typical visualization and editing functionalities such as zooming, rotating, translating, clipping, and selecting and deleting subsets of the data. However, raw datasets require further processing to add value to the data and therefore post-processing functionalities, in addition to basic visualization and editing functionalities, offer more advanced computational tools, see Figure 2. They enable the creation of as-built models of the scanned scene, calculation of geometrical attributes of the model (e.g., area, volume, slope), determination of the deviations of the as-built model from the as-designed model, and detection of clashes. These high-level tasks require other sub-tasks such as data segmentation, fitting geometric primitives, display of modeled surfaces with overlaid point clouds, and the ultimate performance of these packages depends on how those sub-tasks are solved (e.g., at what level of automation and processing speed). Different software packages support these capabilities to different extents.

Software packages for 3DI instruments started out as software that was developed by the instrument manufacturer. These packages were used mainly to control the instrument and to acquire data from that instrument. Since 3D imagining was a new technology for construction and manufacturing applications, there were no software packages that were developed to make full use of the 3D data and that could easily handle the large number of data points collected by a 3DI system. Thus, as the use of and the applications for 3DI systems expanded, this necessitated the evolution of the control software to include additional capabilities such as data processing, modeling, and visualization and then to the development of standalone software to specifically utilize 3D imaging data. As a result, 3DI software can be grouped into three general categories:

1. Software that controls the 3D imaging system
2. Software that uses and processes the 3D data
3. Software that does both (1) and (2)

In general, the software in category 1 is easier to learn than software in the latter two categories. The training required for the software in categories 2 and 3 is similar to that for CAD/CAM packages. The software in categories 2 and 3 can also be grouped by:

1. Sector: manufacturing or construction/civil
2. Specific area: e.g., transportation, structural/architectural, reverse engineering.

In general, software designed for manufacturing and reverse engineering applications differs from those designed for construction or transportation applications in that the input data for the former type of software is less noisy and the required accuracies for the data are higher. Also, the point clouds are smaller for manufacturing and reverse engineering applications. The need

74

for particular software functionality may also depend on the type of application and instrument used to acquire data. For example, for most of the construction applications, medium to long-range instruments are used. The scanning process requires placing an instrument in many locations to get complete coverage of the scene (see Figure 1). In this case, registration of many point clouds to a common coordinate system is required and post-processing software should provide this functionality. On the other hand, point clouds for inspection in manufacturing applications are often acquired with short range instruments mounted on mechanical arms which tracks the locations and orientation of the scanners. These types of systems provide one point cloud where many subsets of data obtained from different instrument positions are automatically registered and the post-processing software does not have to deal with registration or errors resulting from the registration process. Also, inspection tasks often involves known objects or models of the object and thus object recognition is not a requirement for the post-processing software.

Software developed for a specific application has features that make the workflow easier for the user. For example, software for transportation applications has features that incorporates procedures similar to those that would be performed by a surveyor.

3.3.2 Section Organization

The focus of this section is on software developed for terrestrial systems and does not include software for airborne systems.

An online magazine, Point of Beginning (POB), conducts annual surveys of 3DI hardware and software. The surveys are very comprehensive, but the surveys are focused on software and hardware for construction applications. The list of software in the POB survey is given in Appendix C, Table C. 1. This list also includes software packages for 3D imaging systems that are not in the POB survey. The link to both POB surveys is: http://laser.jadaproductions.net/ (last accessed 7/21/10). The questions from the POB survey were sent to some of the software companies not in the 2010 POB survey and their responses are given in Table C. 2.

The POB software survey data was available from 2004 to 2010 (data from the 2008 survey is not available). The number of software packages from 2004 to 2010 is shown in Figure 58.

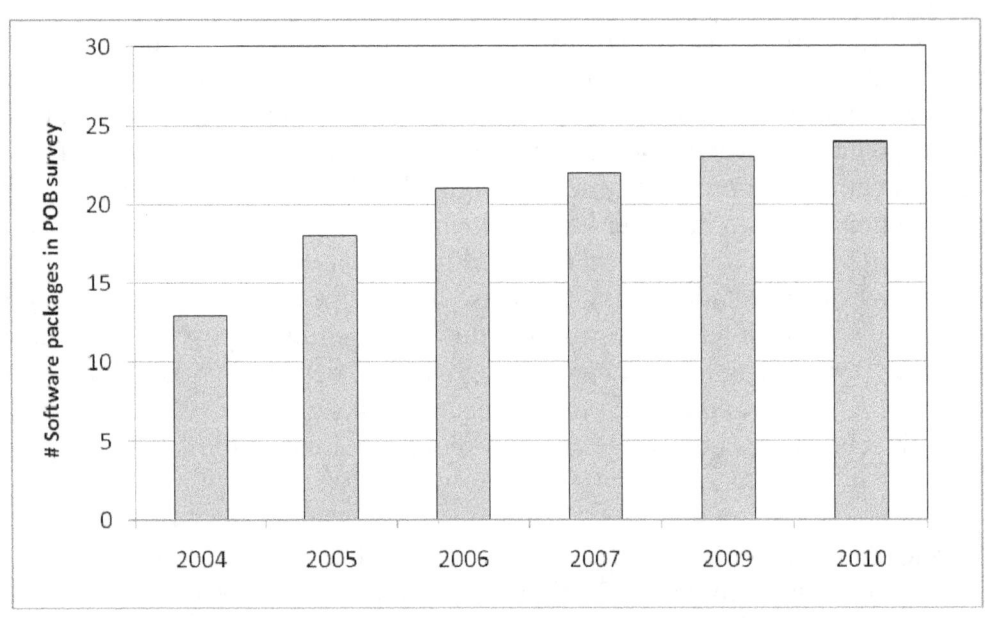

**Figure 58. Number of software packages in the 2004 to 2010 POB survey.
(Data from 2008 not available).**

One method of analyzing the trends in software was to examine how the responses to the POB survey questions changed over a period of time. For example, an increase in the number of "Yes" responses to the question "Automatic removal of noise (e.g., cars, vegetation)? Yes/No" over time would indicate improvements to the software and a market need as software improvements are usually driven by customer demands or needs. This method, however, yielded counterintuitive results most of the times. The main reasons for this are 1) subjectivity - different interpretations of the survey questions and/or potential biases in the responses, and 2) software packages available in earlier surveys were not available in later surveys and vice versa.

The intent of this section is not to reproduce the POB survey, but to describe the capabilities of the software for 3D imaging systems. With this intent, a taxonomy of the software features or capabilities was developed; the major groupings in the taxonomy for software features and capabilities are: Platform, Data Input / Output, Point Cloud Processing, Modeling, Analysis, and Visualization

The following sections describe each of these groupings in more detail. There is no attempt to rate the software packages in any way.

3.3.3 Platform

All of the commercial software packages operate in a Windows environment and do not require a high-end workstation. The workstation requirements are typical for a moderate to professional level workstation: CPU of about 2.5 GHz, RAM range from 512 MB to 2GB, hard disk space of about 2 GB to 4 GB, and typically an accelerated 3D graphics card with support for Open GL

76

and a minimum of 128 MB of memory. However, a workstation with faster and multiple CPUs, higher RAM, and a better graphics card could improve efficiency.

The software packages can run on a 64-bit machine, but most of the software is not written to take full advantage of a 64-bit machine. However, more software is being written to do so. Approximately half of the software packages listed in Appendix C support the client-server architecture to allow multiple users to share a floating license. A useful feature of the client-server architecture is the ability for multiple users to work on the same project simultaneously, and about 50 % of the packages offer this feature.

3.3.4 Data Input / Output

Interoperability between software packages is a very important consideration. This includes the ability to:

1) Import files from
    a. Different 3D imaging packages
    b. CAD package into a 3D imaging software package
    c. Surveying packages (e.g., control data)

2) Export files to
    a. Another 3D imaging software package
    b. From a 3D imaging software package to a CAD package
    c. Specialized software packages (e.g., slope stability software packages)

Another consideration is other data that are imported or exported along with 3D data (e.g., intensity, color, digital photo) and what metadata (e.g., name of 3DI system, measurement units) are supported.

In general, software that is primarily used to control an instrument is developed by the instrument manufacturer. This type of software can import data (point clouds) in its native format, in ASCII format, and in a very limited number of file formats from other instruments. These software packages can generally export point cloud data in their native format and in ASCII format.

Software that processes and utilizes data from 3D imaging systems is developed either by the instrument manufacturer or by a third party. This type of software can import data from more instruments and can export data in more formats than software developed to control an instrument. This is especially true of software developed by a third party. Some of the common export file formats are DWG, DXF, VRML, IGES, OBJ, STL, ASCII, and TXT. Other formats that are not as common include DGN, JPEG, TIFF, AVI, CGM, PLY, and LAS.

Interoperability can also be achieved though the incorporation of the functionality of one software package into another software package via plug-ins. For example, some point cloud processing functionality is made available from within a CAD environment and some instrument control functionality is made available from within point cloud processing software. Currently, some software vendors are using this strategy.

3.3.5 Point Cloud Processing

A deliverable from 3DI systems is a point cloud. Point cloud processing consists of the ability to display, process and to manipulate large numbers of points. Most of the software packages claim to be able to load 100 million or more points or up to the limit of the computer hardware. However, some problems encountered when processing and manipulating datasets of 100 million points and larger include slow start up time (i.e., time to load points into memory) and slow response time when viewing or navigating (e.g., rotations, refreshing the view), and crashing of the program.

Some common features for processing and manipulating millions of points include:

- *registering datasets*: Most software packages have the capability to register datasets with 70 % of the packages able to register without the use of targets.

  There are two methods to register datasets: with special targets placed in the scene and without targets. For the first method, about 38 % of software packages can automatically detect these targets while others require manual identification of the targets. Once the corresponding targets are identified, the registration process is straightforward.

  The second method for registration does not require special targets to be placed in the scene. This method of registration (also known as target-free registration) can help improve the 3D imaging workflow. For target-free registration, there are two ways to register. First, an automated registration can be performed but it requires overlapping regions in the datasets and that the datasets are roughly aligned to begin with. The algorithm that is mostly used to perform this type of registration is a form of ICP (Iterative Closest Point) and the required overlap region ranges from 20 % to 50 %. A second method involves manually picking a minimum of three corresponding points in the two datasets.

  Another form of registration is geo-referencing the datasets to a global coordinate system. Most software packages have this capability.

- *automatic detection of scan targets*: Target detection is usually based on shape, intensity, and/or pattern and 38 % of the software packages are able to automatically detect scan targets (targets placed within a scene for registration). This feature helps improve the 3D imaging workflow.

- *clipping or cropping data*: The ability to easily set and change limits within which the data points are visible.

- *segmenting data*: The ability to easily select a of set points. The selection is generally made using a user-defined fence, and the user can select the points within or outside the fence. Another way of segmenting the points is by using a surface such as a plane. The selected points can then be deleted, exported, or further processed. Data segmentation is mainly a manual operation although certain software are able to segment the point cloud by various parameters (such as elevation from a ground plane). The ability to automate this task would greatly improve efficiency.

- *decimating the number of points*: The ability to reduce the number of points in the data set. A common method of reducing the number of points is to delete every other point or every X number of points. This method does not have any "intelligence" associated with the point selection. A better method, but less common method, is to remove more points in a more densely populated area and fewer points in a less populated area or to keep more points in regions where there are more variations in the surface (e.g., break lines, discontinuities in the surface, changes in the slope). This feature is useful in removing redundant points in regions where point clouds overlap.

- *filtering points*: This feature is found primarily in software packages that mainly control an instrument where the collected points may be filtered based on signal strength or other criteria. About 70 % of the software packages have algorithms to remove unwanted points such as those from cars, vegetation, etc.

- *merging of datasets*: In its most basic form, this feature allows the user to import of multiple datasets and to combine them into one. All software packages that can import point clouds have this capability. When combining datasets, some areas may become more densely populated as a result and some packages have algorithms to reduce the duplicate points (i.e., points that are close to each other) in these regions.

- double precision processing: This feature is available in all software packages

Other less common features include:

- *batch mode processing*: The ability to write a script to perform a set of tasks. Certain processes can take a long time when millions of points are involved. The ability to perform batch mode processing would increase a user's efficiency; however, this feature is only offered in one package.

- *viewing multiple data sets at the same time and manipulating (translating and rotating) each data set independently*: This capability is very useful for manual registration of two or more data sets.

- *gridding*: Some software packages allows the user to generate a regularly gridded set of points from a point cloud.

- *multiple return analysis*: Multiple returns are caused when a laser beam hits multiple objects separated in range. The ability to detect the multiple returned signals allows the user to detect objects that may otherwise not be visible. This feature is only offered in one software package.

3.3.6 Modeling

Common deliverables derived from 3D imaging data are 2D plans and 3D models such as wireframe, surface, and geometric models. The process to create a 3D model is mostly manual process and can be time consuming depending on the type of model, amount of detail, and the required accuracy. Automated object recognition from a point cloud is the subject of much research and this capability is currently very limited; for example, the development of software that automatically finds and fits geometric primitives such as lines, planes, cylinders, and spheres in point clouds.

Most commercial software packages have the capability of fitting geometric primitives such as lines, planes, spheres, and cylinders; however, this process is not fully automated as evidenced by the responses to the following survey question:

"Is fitting of lines, planes and shapes to cloud done manually or automatically, or both?"
    Both = 59 %
    Some automated (e.g., lines, planes) = 7 %
    Manual = 17 %
    No or N/A = 17 %.

About 35 % of the software packages have the capability to perform automatic extraction of standard shapes (e.g., pipe fittings, structural steel members) from the point cloud. However, the degree of automation is unknown. Only 28 % of packages have the capability of fitting CAD models to the point cloud using standard object tables/catalogs. However, about half of the software packages allow features to be defined with user-created code libraries. A few of the software packages are plug-ins to CAD packages and operate in a CAD environment.

About half of the software packages offer features such as solid modeling, edge detection, and automatic generation of a polygonal mesh. A polygonal mesh that is commonly used is a triangulated irregular network or a TIN. A common use of a polygonal mesh is the generation of a topographic map. The detection of breaklines in the topographic map is not a common feature in the software packages (this feature is offered in a few software packages that processes data from airborne systems).

Other features such as the ability to automatically track lines or automatically calculate centerlines of shapes (e.g., pipes) are offered by about half the software packages. These features are somewhat application specific. For example, the ability to track lines or follow edges would be a feature in software developed mainly for transportation applications.

Another type of 3D model that is currently not as common is a BIM (building information model). "A BIM is a digital representation of physical and functional characteristics of a facility" [76]. A BIM allows different stakeholders at different phases of the life cycle of a facility to add, extract, update or modify information. The information could include material properties, equipment manufacturer, date of purchase, and date of last maintenance. Therefore, the import and export of a BIM model will have to support metadata (e.g., facility management database).

3.3.7 Analysis

The software packages offer the basic features for analyzing the 3D data such as the measurement of:

- distances between
    - two points
    - points and object/model
    - two models
- angles between
    - two lines
    - line and plane
    - two planes
- attributes of fitted geometric primitives such as pipe diameter, pipe length, and sphere center.

About 50 % of the software packages have the ability to detect measurement outliers based on fitting geometric primitives. However, the ability to detect outliers that originate from split signals caused by object edges is not a common feature offered by the software packages.

A useful feature is the ability to calculate volume and this feature is offered in about 50 % of the software packages. Volume calculation depends on the complexity of the surface which defines the volume. For example, the calculation of the volume of surfaces with undercuts is more difficult and may require user intervention. Another potential solution is the use of tetrahedralization.

Another feature that is useful for renovation/revamp projects is interference or clash detection. This feature is offered by about 75 % of the software packages and the need for this feature appears to be increasing. The degree of automation and sophistication of the clash detection varies between different packages.

Another use of 3D imaging data is for inspection where the acquired data (point cloud or as-built model) is compared to plans or models (as-designed). About 65 % of the software packages have the capability of comparing the point cloud with floor plans and engineering drawings of objects. Packages with this capability also have the capability of reporting (text and/or graphically) the differences between the as-built and the as-designed models.

In the POB survey, 65 % of the responses to the question "Create statistical quality assurance reports on the modeled objects? Yes/No" was "Yes". However, another important piece of information that is needed is the uncertainty of the fitted parameters such as the uncertainty of a fitted sphere center or the orientation of the axis of a fitted cylinder. It is important because the uncertainty of the model parameters (e.g., geometric dimension) has to be known in order to determine with a given level of confidence whether the model is within tolerance or not. Only a few of the software packages provide the uncertainties of the fitted parameters.

Slope analysis is not a common feature offered in the software packages as it is specific to construction applications (e.g., mining, excavation, roadway surveys). This feature is offered in some airborne software packages.

In general, software packages with the capability of registering data sets also generate a report of the registration error. It should be understood that the average distance between corresponding points within the overlap region used for registration is not larger than the registration error. However, average distance between corresponding points outside of the overlap region used in the registration may be larger.

3.3.8 Visualization

An important benefit of 3D imaging data is the ability to visualize a scene, especially if the scene is very complex. The ability to view the scene from any angle and to zoom in to view more details is a powerful tool. The basic navigation capabilities offered by all of the software packages include pan, tilt, and zoom. A majority of the software packages (70 %) have an intelligent level-of-detail display based on scale of view. Other less common features that aid in scene viewing/navigation include the ability to view the data from the point-of-view of the instrument or from a user-selected point-of-view.

Coloring of the points in a point cloud is another key visualization tool. Most software packages offer some or all of the following coloring schemes:

- False coloring
- Coloring based on:
    - range
    - intensity
    - true color - RGB obtained from 3D imaging system
    - true color - RGB from photo overlay (not obtained by a 3D imaging system)
    - user specified (e.g., color based on elevation or x-axis values)

In addition, about 70 % of the software packages have the capability to generate texture-mapped models or point clouds for a more realistic appearance. Some of the packages allow the user to capture an image for use in reports or presentations.

Walk-through or fly-through movies based on the point clouds, models or both, can be useful tools when preparing bids or for presentations to clients. This feature is supported by 88 % of the software packages. The ease-of-use varies between packages.

Most software packages have the capability to make profiles and cross sections. About 45 % of the software packages can generate contours.

Some software packages offer a free "viewer" which is useful as it allows others to view the point cloud or model without having to purchase the software.

4 Future Trends and Needs

4.1 Hardware Trends and Needs

4.1.1 Active optical 3D imaging needs in manufacturing

In 2007, the Manufacturing Engineering Lab (MEL) at NIST investigated the needs and requirements of active 3D imaging technology for automation in assembly and mobility applications in the manufacturing industry. The task started with a review of recently published documents [77-83] describing the future technology needs and challenges for developing and applying next generation robots to many new social and industrial tasks in the 21st Century. During the review particular interest was placed on collecting data describing expected 3D imaging requirements in next generation manufacturing applications. The material collected in the review process was then used in the preparation of a list of survey questions and distributed to a selected group of company representatives to gather their expert opinions on the topic. The survey was sent to representatives in the following industries: aircraft manufacturing, automotive manufacturing, manufacturing inspection and metrology services, robots for assembly, and marine technologies. The representatives provided expert insights on manufacturing applications and the corresponding requirements needed for industry acceptance.

These were the major application areas suggested by the survey participants:

- Safety systems
  - Collision avoidance for robots, Automatic Guided Vehicles (AGVs) and large assembly parts that are in motion
  - Supervisory safety systems on the assembly line for human and robot cooperation
- Automatic material handling systems
  - Part identification and tracking
  - Part acquisition and manipulation
  - For elimination of errors in part selection for assembly
- As-built modeling and inspection of parts and assemblies
  - Provide corrections to manufacturing equipment and for refinement of parts for assembly
  - Reverse-Engineering – precise comparison of as-built with design specifications
- Optimizing the distribution of tasks between robots and humans in cooperative assembly tasks

The following is a summary of the 3D sensing requirements that were provided by the survey participants:

- Need performance standards and tests to characterize measurement performance on static and dynamically moving parts – edges and corners, different materials and colors, different target angles

- 3D measurement requirements for tooling and inspection of parts
  - General: ± 0.1 mm uncertainty in three dimensions
  - For precision Reverse-Engineering: ± 0.0025 mm uncertainty in three dimensions
  - Frame rates of 1000 Hz or higher
  - Ranges of up to 10 m or more
- 3D measurement requirements for assembly line operations (for static and moving parts)
  - Macro Level: ± 1 cm uncertainty in three dimensions
  - Micro Level: ± 1 mm or less uncertainty in three dimensions
  - Micro-Micro Level: ± 0.1 mm or less uncertainty in three dimensions
  - Frame rates between 15 Hz and 30 Hz
  - Part tracking at speeds of (3 to 4) m/min
  - Compensation for perturbations (1/2 G forces are possible)
- Performance measures and analysis to evaluate repeatability and reliability
- Performance measures and standards for sensor interfaces

Based on the data collected in this study, it appears that companies in the manufacturing industry are only beginning to understand the capabilities and performance that can be provided by the next generation active optical 3D imaging technology. They are testing available products and prototypes to establish some baseline understanding of the performance characteristics. All of the study participants pointed out that performance measures and standards are needed for this technology to be accepted and considered for application on the shop floor.

The standards and performance issues of next generation optical 3D perception systems were recently discussed at a workshop on Dynamic Perception: Requirements and Standards for Advanced Manufacturing on June 11, 2009 in conjunction with the International Robots, Vision & Motion Control Show in Rosemont, Illinois. The complete report on the dynamic perception workshop is available at: http://www.nist.gov/manuscript-publication-search.cfm?pub_id=904612 . The workshop addressed the following four main questions regarding the needs and steps necessary to advance the use of dynamic 3D imaging technology in advanced manufacturing robotics and machine/assembly automation [84]:

- Are there sets of manufacturing scenarios or tasks that can be identified as candidates for employing sensing to advance robotic capabilities?
- Which of these scenarios or tasks are the "low hanging fruit" that can have a near-term impact?
- What expanded capabilities and performance are needed to advance perception systems in manufacturing robotics?
- What are essential metrics for evaluating perception systems in these scenarios?

There was considerable discussion regarding the first three items on the list and a very interesting observation made by the workshop organizers:

"The participants generally identified material handling tasks as the best first priority for candidate scenarios. During the initial phase in which participants individually listed possible

scenarios, only one participant mentioned part processing such as machining, welding, cutting and painting while one other mentioned part joining. Other participants listed variations on the perception of objects for a broad range of picking and placing operations, and the open discussion centered on defining and refining pick and place scenarios. We concluded from this that perception during transformative operations, like cutting, welding or milling, is a lower priority at this time, and that initial success at perceiving objects for grasping and placement would be a natural first step to loading and handling objects during processing."

There was insufficient time for the workshop participants to get into discussing details of metrics for evaluating sensor performance in the selected scenarios. At the end of the workshop, discussion was initiated on the following topics: requirements for possible standard artifacts or targets, measurement resolution and accuracy, and on visual characteristics (such as reflectance, emissivity and occlusion of targets) that can affect sensor performance. These topics were to be pursued at post-workshop meetings.

4.1.2 NIST program in measurement science for intelligent manufacturing

In the fall of 2007, the Manufacturing Engineering Lab at NIST initiated a new program in Measurement Science for Intelligent Manufacturing Robotics and Automation which addresses several goals identified during the NIST smart assembly workshop in 2006. A complete report on the smart assembly workshop can be obtained at the following website link: http://smartassembly.wikispaces.com/ . The list of goals includes: monitoring shop floor activities to maintain virtual models, improving sensing capabilities, providing additional capabilities and adaptability in assembly systems. Performance measures and standards are pointed out as being central to effective in-process measurement and for real-time continuous metrology in the manufacturing process. As a follow-up to the 2006 workshop, a workshop on requirements and standards for dynamic perception in advanced manufacturing was held in 2009 [84]. In response to the needs expressed by industry participants, MEL initiated a program in Intelligent Manufacturing. The main challenge of the program is to develop robust dimensional metrology methods in order to evaluate the performance of these new sensors for manufacturing in dynamic unstructured environments where people and machines interact. This is called out as an important area of innovation in the NIST assessment of the United States Measurement Systems (USMS) needs and particularly in Laser-based 3D Imaging Systems [85].

The following manufacturing industry needs are being addressed by the MEL program in measurement science for intelligent manufacturing systems (http://www.nist.gov/el/isd/si/msimra.cfm):

- New and reliable inexpensive safety systems to protect humans and avert damage to equipment in dynamic shop floor environments
- Flexible and adaptable control systems that can adjust to variations in the shop floor environment and in materials
- Advanced sensors to locate and identify parts and which can determine properties and manipulate parts in 3D space

- Adaptable, safe and repeatable simulation environments that interact seamlessly with real-world manufacturing equipment and environment
- Sensors and measurement technology to measure motion and displacements in nanomanufacturing and in nanorobotics
- Advanced control and positioning systems for 3D micro/nano structures and devices

The performance requirements and standards being addressed by the program for advanced perception and measurement systems include:

- Performance test methods, characteristics and metrics for static and dynamic 6 degree-of-freedom (DOF) sensing systems
- Standards which describe sensor system product performance and help match products to applications
- Calibration procedures for 2D, 3D and 6 DOF sensors to enable cross-sensor comparison and evaluation

4.1.3 Operational requirements of active 3D imaging systems for mobility applications

The U.S. Department of Defense (DOD) has established programs for deployment of ground robotic systems in future combat operations. These encompass large manned and autonomous Unmanned Ground Vehicles (UGVs) as well as smaller UGVs (PackBot size vehicles) performing various tactical mission scenarios. Possible tactical missions for UGVs include: reconnaissance, active or passive surveillance, communication relay, mine detection and clearing, targeting, search and rescue, supply, terrain control/denial, forward observation, and lethal or non-lethal missions. These missions will require the vehicles to drive autonomously in structured and unstructured environments which could contain traffic, obstacles, military personnel as well as pedestrians. UGVs must therefore be able to detect, recognize and track objects and terrain features in very cluttered environments. Laser-based 3D imaging sensors (laser radar) has demonstrated the ability to provide reliable real-time 3D imaging with sufficient resolution to effectively model the 3D environment at distances out to 100 m for on and off road driving. Although advances have been made over the last 5 years, and robust performance has been demonstrated, the relatively high cost of these sensors is still an issue which needs to be addressed.

An overview of the sensor requirements for driving UGVs is presented in a book published by NIST in 2006 on Intelligent Vehicle Systems [55]. Chapter 7 of the book is on Advanced LADAR for Driving Unmanned Ground Vehicles. Since the sensor operational requirements are still valid for UGVs, the material from the book will be used extensively in this report and updated only if necessary. The use of laser radar for real-time active 3D imaging is also being considered for autonomous navigation, mapping and collision avoidance on small UGVs (such as the iRobot SUGV and PackBot and the Foster Miller Talon). The operator would just need to provide a command to the robot to go to a designated target location. The sensor could also be used for imaging and mapping of terrain (streets and buildings) in urban settings as well as unstructured environments (such as caves). Because of the limited size and available power on small PackBot size robots, baseline sensor performance requirements will need to be adjusted.

Real-time active 3D imaging technology is also beginning to be used in automotive safety systems and in automated material handling systems used in manufacturing. New developments in sensor performance requirements in these applications will be presented.
The sensor operational requirements are broken out into the following categories:

1. Autonomous on and off road driving for UGVs with active real-time 3D imaging
2. Autonomous navigation and perception for small PackBot size robots
3. Automotive safety systems
4. Autonomous navigations and collision avoidance for industrial Automated Guided Vehicles (AGVs)

These requirements will be presented in detail in the following sections.

4.1.3.1 <u>Operational requirements of laser radar as a real-time 3D imaging sensor for driving UGVs (Unmanned Ground Vehicles).</u>

An initial set of baseline LADAR requirements for driving UGVs was established by NIST in 2002. These were based on the experience NIST gained from participating in the Army Demo III program which investigated the use of LADAR for on and off-road autonomous driving. The original baseline specifications are described in section 2 of Chapter 7 in [55].

"NIST envisioned the need for two types of LADAR range imaging sensors for this type of application - one having a wide FOV (40° x 90°) with an angular resolution of about 0.25° or better per pixel, and the second a foveal LADAR having a narrow FOV of approximately 1/10th of the wide FOV with an angular resolution of about 0.05° or better per pixel. The intent was to make the foveal LADAR quickly steerable to points-of-interest positions within the wide peripheral angle FOV LADAR at a rate of at least 3 saccades (point-to-point moves) per second. Both types of LADAR sensors were expected to have an uncertainty of about ±5 cm or better in range, and be able to detect the ground plane out to a distance of better than 50 m and vertical surfaces out to a range of at least 100 m. Frame rates of higher than 10 Hz were required. Both types of LADAR were expected to be eye safe and be provided with the capability of penetrating dust, fog, grass and light foliage (either by sensing multiple returns or looking for the last return), and be able to operate in full sunlight conditions. Small size and low cost were also emphasized as important requirements."

Detailed requirement specifications for LADAR were published in a 2002 NIST BAA (Broad Agency Announcement) titled "Next Generation LADAR for Driving Unmanned Ground Vehicles." These are listed in Appendix B of [1]. In 2005, NIST updated the above requirements based on guidance from DARPA and on the outcome of an ARL funded project to establish perception and autonomous driving requirements for a tactical Road Reconnaissance mission. The updates to the baseline requirements, which were presented in section 2.3 of Chapter 7 in [55], are as follows:

"Initial study results have indicated that the original NIST BAA requirements still stand but need some minor changes and additions. Figure 59 is a conceptual diagram of a single LADAR

sensor or dual sensors intended for autonomous UGV driving needs. A Wide FOV (WFOV) (40° x 120°), coarse angular resolution (0.25°) LADAR is needed for peripheral vision and a Narrow FOV (NFOV) (4° x 12°), fine angular resolution (0.025° or better) is needed for saccadic foveal perception. The intent is to steer the foveal LADAR to areas of interest within the field-of-regard of the peripheral LADAR sensor at a rate of 3 to 10 saccades per second. The higher resolution is necessary to detect, classify and track objects and personnel on or near the path taken by the vehicle at distances up to 200 m, when the vehicle is operating at top speed. Some sort of image stabilization or image motion compensation must be provided for the high resolution camera."

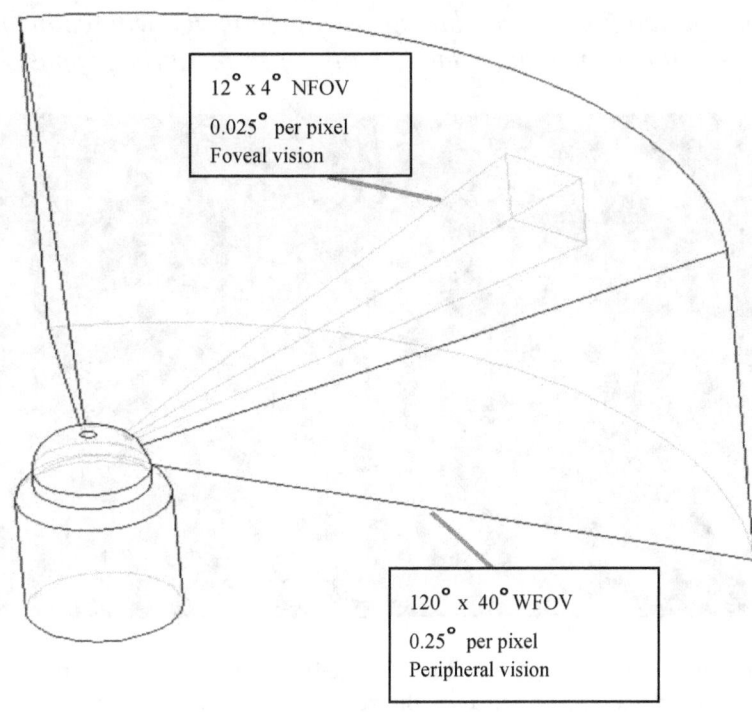

Figure 59. NIST conceptual diagram of a single 3D imaging sensor or dual sensors for UGV autonomous driving needs.

The 2005 updates also included the following changes and additions to the LADAR requirements:

1. Range uncertainty (standard deviation plus bias) must be ± 5 cm
2. Range resolution should be at least 15 cm or better
3. Provide intensity and color data for improved object classification and recognition.
4. Include: Concertina wire detection, thin wire and object detection, and detection of rocks hidden in grass and foliage. Publications [86, 87] describe research which was conducted in this topic area.

Additional LADAR performance requirements for increased safety in high speed UGV driving were generated by NIST in a study conducted for ARL. The following text came from [55]:

"As UGVs near deployment in military operations, there is a growing need for high speed driving safety. In a study conducted for ARL, NIST developed some initial perception performance requirements for 3D imaging systems at distances out to 100 m. The 3D imaging system must be able to detect and identify a person in the path of the vehicle in time for the vehicle to stop or avoid hitting the person. Tests conducted at NIST with a long range, variable resolution, high performance 3D imaging system have concluded that an angular resolution (0.2° to 0.25°) enabled the detection of objects the size of a human at 100 m, but not the identification of them. When the angular resolution was set to 0.02° (foveal LADAR perception), it was possible to apply segmentation approaches to identify a person as shown in Figure 60. There are 600 pixels on the target at 100 m. In addition, by combining the range image with color, fewer than 100 range pixels on target may be required to identify a person at ranges past 100 m."

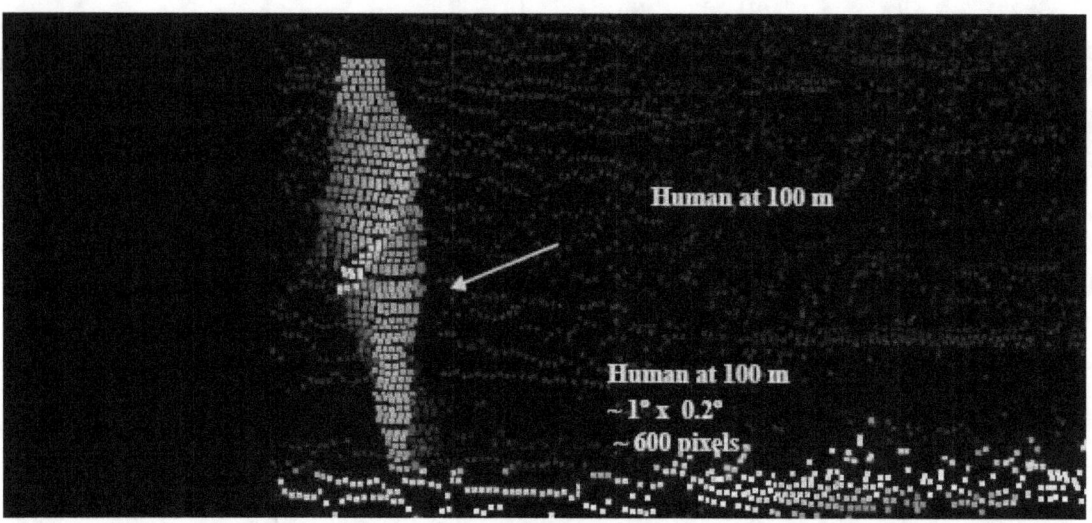

Figure 60. Recognition of a human target at 100 m. Range image was obtained with an angular resolution of 0.02° per pixel. This is approximately the resolution of unaided human foveal vision.

In January of 2009, the ARL Robotics Collaborative Technology Alliance (CTA) conducted an assessment and evaluation of multiple algorithms for real-time detection and tracking of pedestrians using LADAR and vision sensor data. A robotic vehicle was equipped with two pairs of stereo cameras, GDRS Gen III LADARs, and SICK line scanner imagers. The performance measures for static and dynamic multiple human detection and tracking from a moving vehicle are presented in [88]. A robust and accurate independent pedestrian tracking system, developed by NIST, was used to provide ground truth. The variables in the test scenario included robot vehicle speeds of (15 and 30) km/h and pedestrian speeds of (1.5 and 3.0) m/s. The performance evaluation was intended to provide evaluation of sensor and algorithm performance for Military Operations in Urban Terrain (MOUT). Humans were detected and tracked reliably at distances well over 50 m. The assessment was successful in validating the measurement system and procedures for evaluating sensor and algorithm performance. Additional information on the assessment approach can be found in [89].

4.1.3.2 Baseline requirements for autonomous navigation and perception on small UGVs (such as PackBots)

The following minimum 3D imaging performance requirements for small robots were provided by ARL in the presentation material at the 2010 SPIE Defense & Security conference [40].

- Frame rate: better than 5 Hz (10 Hz to 12Hz desirable)
- Pixels per frame: at least 256 (h) X 128 (v)
- FOV: at least 60° (h) X 30° (v)
- Spatial resolution: at least 0.25°/pixel
- Range (DOF): at least 20 m (longer ranges desirable)
- Range bin sampling interval: 10 cm or better
- Range uncertainty: < ± 2 cm
- Laser wavelength: eye safe (e.g. 1550 nm)
- Sensor power requirement: < 30 W
- Sensor weight: ≈ 1 kg
- Intensity image: for overlay on range images
- Must operate in bright outdoor ambient light conditions

The research at ARL is driven by the need to develop low cost, compact, low-power LADAR imagers for small UGVs for navigation, obstacle detection and avoidance, and target detection and identification.

4.1.3.3 Desired sensor measurement attributes for automotive safety systems

The transportation industry and the Government are looking at advanced technology to improve vehicle safety and reduce traffic accidents [55]. In 2005, the Department of Transportation (DOT) initiated a new program to develop and evaluate an Integrated Vehicle Based Safety System (IVBSS) designed to provide warnings for imminent rear-end, lane-change, and road departure crashes on light vehicles and heavy commercial trucks. Information about the program is available at the following website: http://www.its.dot.gov/ivbss/index.htm .

Figure 61 is an illustration, provided by a commercial 3DI system developer, of what automotive safety and driver assist applications may be possible with an advanced sensor package. A critical component of the IVBSS test program was for NIST to develop verification test procedures and an independent measurement system (IMS) for determining acceptable performance and study the characteristics of the prototype warning system on public roads. After studying the problem, NIST decided to look for a 3DI system which had the following measurement attributes:

1. Determine location of obstacles (range and azimuth)
2. Determine size of obstacles
3. Determine range rates to obstacles
4. Perform measurements at maximum highway speeds (120 km/h)
5. Detect obstacles at far range (> 65 m)
6. Sufficient FOV to view and detect obstacles in the path of the vehicle

a. rear-end sensing – FOV must cover the road in front to measure curvature of the road
b. road-departure sensing – FOV must cover the shoulder in the forward direction of the vehicle and take into account the curvature of the road and obstacles directly to the side (e.g., jersey barriers)
c. lane-change sensing – FOV must cover the adjacent lane directly to the side of the vehicle and possibly to the rear to detect passing vehicles

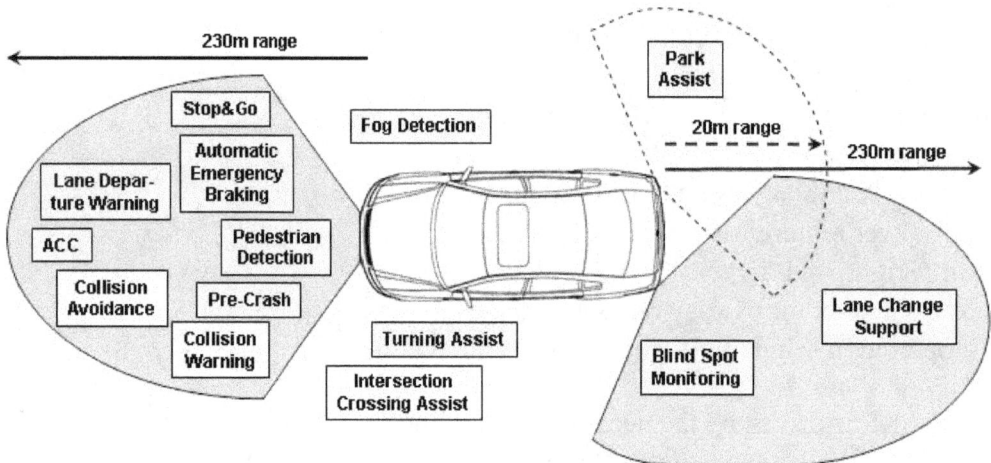

Figure 61. Application of a 3D imaging system for automotive safety and driver assist systems (courtesy of IBEO Automotive Sensor GmbH).

NIST decided to use laser-based 3DI systems to measure ranges to obstacles in front and to the sides of the test vehicles. The complete IMS system description along with measurement procedures and system performance is available in [90]. Figure 62 illustrates the IMS systems FOV capability.

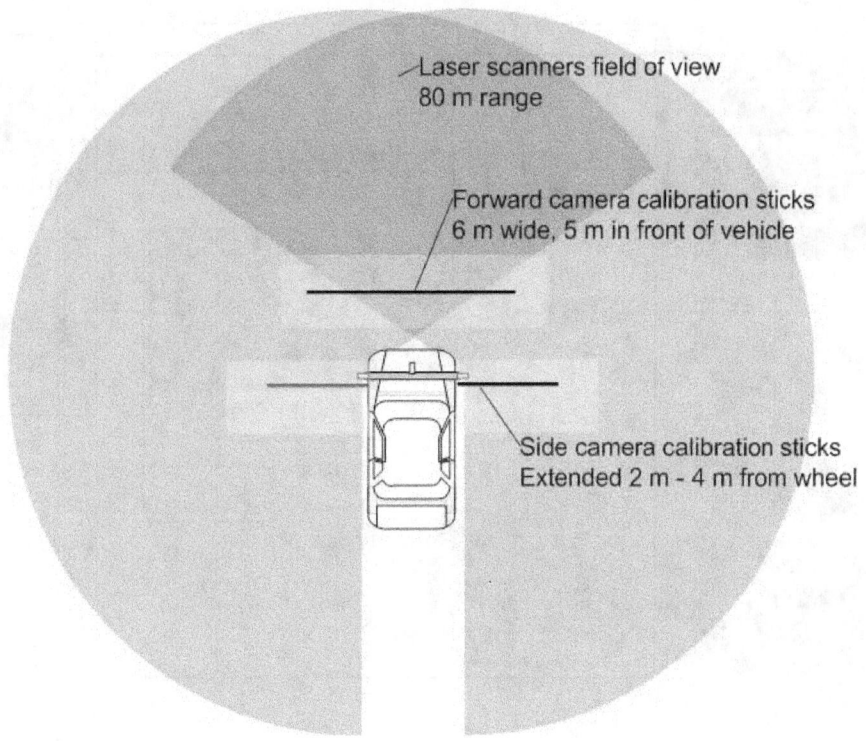

Figure 62. Measurement system FOV [40].

"Two commercially developed real-time laser scanners were mounted on the front corners of the vehicle to provide distances to objects around the test vehicle at ranges from 1 m to 80 m. Each laser scanner uses four fixed lasers mounted to provide a 4° vertical field-of-view. A rotating mirror scans the laser beams a 320° horizontal field-of-view. The laser scanner includes an electronic control unit (ECU) that fuses the range data from both scanners into a single Cartesian coordinate system." [90].

Figure 63 shows all the system components of the IMS system. The sensor calibration procedures, data collection methods, system measurement procedures, and system acceptance tests are described in [90]. Figure 64 shows the view of the laser scanner data display during a test run.

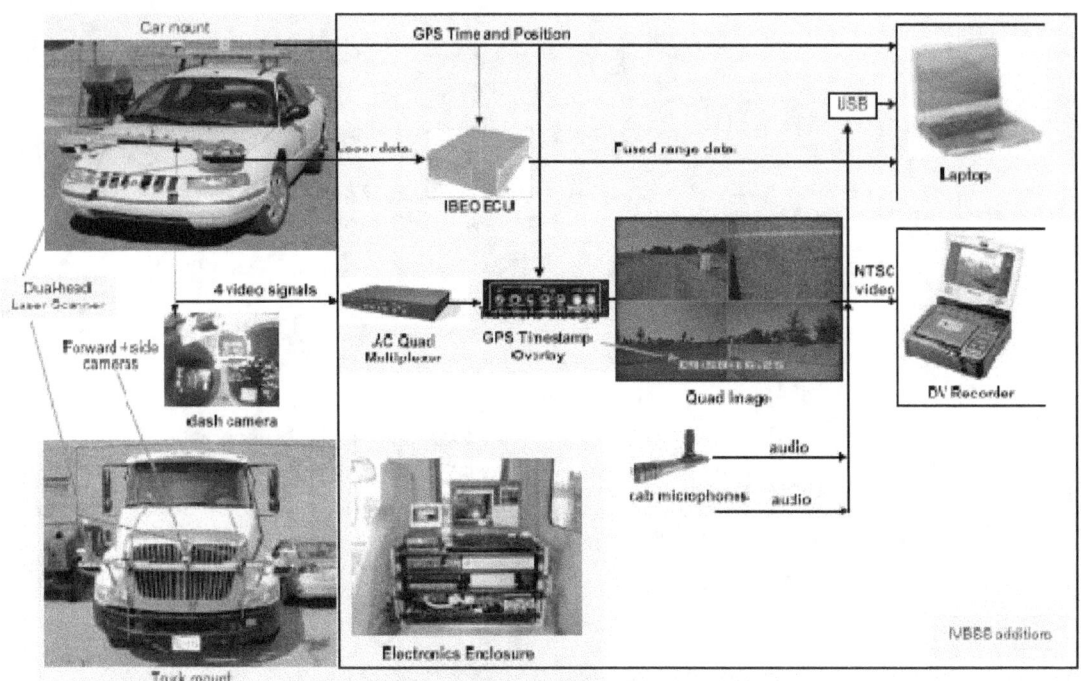

Figure 63. Measurement system components [90].

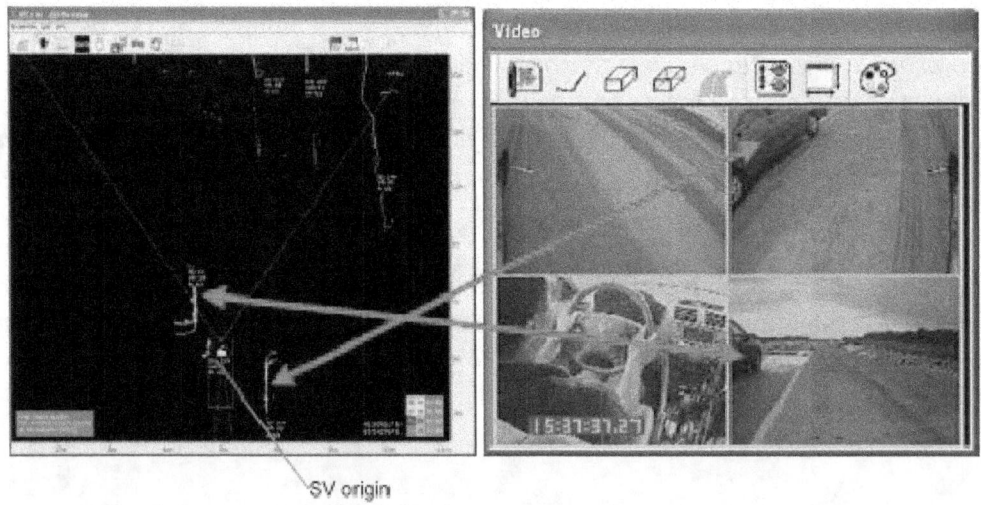

Figure 64. View of 3D imaging data collected during a test run [90].

The IMS contributed to the IVBSS program by providing vehicle safety system performance validation data and measurement capabilities for dynamic on-road testing. This contribution was summarized in [90] as follows:

"The independent measurement system developed by NIST is a real-time, vehicle-based system for measuring range and range-rate to objects surrounding a test vehicle and for boundaries. IMS users can post-process and analyze the test data to achieve a high-

94

degree of confidence in its accuracy and reliability. The system's forward-range uncertainty is approximately ± 1 m at distances up to 60 m and at target closing speeds of 20 m/s.

Independent measurements support the user in:

- *Identifying errors in warning range*
- *Identifying warning system latencies*
- *Identifying errors in data time-stamping*
- *Modeling errors and applying the models to compensate for range and timing-delay errors*

The system provides a wide range of measurement and data collection capabilities for both on-track and on-road testing without the need for instrumentation on other vehicles or of the roadway. A less-expensive approach such as using reference marks painted on the track surface or a calibrated forward-looking camera may suffice as an alternative or as a backup check for IMS malfunction."

4.1.3.4 Autonomous navigation and collision avoidance for industrial Automated Guided Vehicles (AGV)

AGV technology needs were evaluated in a study [83] contracted by NIST . As part of the study, a set of questions was sent to 26 companies. Even though only eight responded to the survey, they represented most of the major AGV vendors in the U.S.

The following were the questions in the survey:

1) What technological gains can significantly improve AGV operational capacity, expanding and opening up new markets for AGV vendors?
2) Specifically regarding operational speed, what improvements in sensing / machine intelligence does it take to increase speed of operations?
3) Regarding operating in unstructured environments, what are the technology enablers? Which machine intelligent factors must be enhanced? On the applications side, what type of industrial environments would benefit most?
4) What are new task areas for AGVs? In what operations currently performed manually could "smarter" AGVs be used?
5) Improvements in mapping and simulation capability -- within existing facilities, is there a need/benefit to produce facility maps using only a sensor-laden vehicle? If so, what kind of detail and tolerances are required in these maps?
6) For potential AGV users assessing AGV operations, how useful is it to provide rapid virtual representation of plant operation to assess AGV alternatives?
7) What are fundamental cost factors that could be addressed through advanced technology development?

8) Are there industries in which AGVs could be natural operational mode, but they are not being used? If so, are there technical barriers to be overcome in perception and/or control?

9) NIST is considering establishing an informal steering committee to define objectives, assign priorities, and evaluate progress in their AGV program. Also a seminar/workshop is tentatively planned for 2006. Would you be potentially interested in participating in these types of activities?

Reducing costs and developing AGVs which are smart enough to automatically handle obstacles and navigate in unstructured environments were the main areas of expressed interest. Lower costs could be achieved by:

- lower cost sensors
- moving from centrally controlled systems to distributed processing
- operating at higher speeds

Autonomous operation in dynamic unstructured environments could be achieved by:

- improvements in sensor technology
- improvements in perception algorithms
- use of dynamic simulation environments to evaluate operation of autonomous
- AGVs.

The study [83] concluded that 3D imaging sensors (both laser radar and stereo vision) are of greatest interest to the AGV vendor community in providing autonomous operation in dynamic unstructured environments.

As stated in [55], "*One way of increasing cost efficiency in the manufacturing and material handling industries is to increase the speeds that AGVs operate at. Vehicle speeds for indoor operation as high as 2 m/s are being targeted. This, however, increases the possibility of accidental collisions with personnel and other stationary or mobile equipment. The most widely used AGV safety systems use contact bumpers, however, this may not be sufficient to prevent injury or damage at these higher operating speeds.*"

A better approach for improving AGV safety may be to use real-time active 3DI systems as non-contact safety bumpers for collision avoidance. In 2005, the ASME B56.6 bumper safety standard was revised to include non-contact safety sensors. This greatly stimulated the use of laser line scanners and other active 3DI systems on AGVs for improved autonomous navigation and safety.

The next step being taken by the ANSI/ITSDF (new organization replacing ASME) B56.6 standards committee is to develop performance standards for Object Detection Devices and Controls for Driverless, Automated Guided Industrial Vehicles and Automated Functions of Manned Industrial Vehicles. NIST has proposed the use of test pieces to evaluate sensor performance for detecting objects or people (positioned within the contour area of the vehicle) with the vehicle traveling at speeds up to 100 % of vehicle maximum speed. The sensors will be required to detect various standard size test pieces (not yet accepted) having a specified low surface reflectivity and dark optical density. A ballot on these recommended additions is expected in the latter part of 2010.

4.1.4 Construction Trends and Needs

As indicated by the improvements to 3DI systems over the past several years, hardware trends for 3DI systems used for construction applications are faster data acquisition, more accurate measurements, increased resolution, and reduced size or increased portability of 3DI systems.

Also, mobile 3DI systems are a rapidly growing trend as indicated by the increasing number of these systems that have come on the market. These systems are used for city modeling "the fastest growing market segment" [91]. Additionally, mobile systems are used for another growing area, roadside/highway asset inventory, for asset management and safety analysis.

3DI systems are becoming a common tool for surveyors and this is indicated by the integration of 3DI systems and optical surveying and the incorporation of more features of a total station in a 3DI system. The features include onboard controls, "tribrach mounting and a laser plummet, GPS and prism attachments, and dual-axis tilt compensation. These features enable surveyor-friendly workflows such as setting up over known points, resectioning, and traversing."[92].

Currently, the use of 3DI systems requires substantial capital investments due to the high costs of the instruments and the software. Additionally, both hardware and software require well-trained personnel. In a survey for airborne systems, the top three (out of 10) barriers to growth of the airborne market as identified by [93]:

- end users were:
    - cost of data
    - availability of experienced analysts
    - cost of hardware/software

- software and hardware industry were:
    - software functionality
    - cost of data
    - cost of hardware/software

These barriers and their ranking are very likely the same for terrestrial systems.

4.2 Software Trends and Needs

4.2.1 Trends and Needs

Applications such as inspection (e.g., health monitoring of structures, parts in an assembly line) and providing feedback for equipment automation or for situational awareness for equipment

97

operators are potential growth fields which would require real-time data acquisition and processing. Additionally, the hardware trend for mobile 3D imaging will benefit from real-time data processing.

The software needs for construction and manufacturing are listed below.

- *Interoperability between software packages and between software and hardware packages.*

  Interoperability between software packages is probably the most important improvement needed. For point cloud data, this issue is being addressed by a standard, Specification of 3D Imaging Data Exchange, written by the ASTM E57.04 Data Interoperability Subcommittee that is currently up for ballot.

  The other types of files are CAD/BIM files, and the seamless exchange of these types of files needs to be improved. It is anticipated that the use of BIM will increase, and interoperability between software packages will become critical. For example, information exchange between software for construction-related applications and for mechanical, HVAC systems.

  As alluded to in Section 4.1.1, airborne imaging is very closely related to terrestrial imaging and interoperability between the software packages for airborne data and terrestrial data would be a beneficial capability.

- *Tools for automatic object recognition, segmentation of planes, spheres, trees, removal of "noise".*

  A higher level of automation of tasks such as object recognition, segmentation, fitting, and data filtering (e.g., noise removal) is one of the more important improvements in the next generation software. Yet, it is also the most challenging task as it requires research and development of robust algorithms. The difficulty in accomplishing this task is evidenced by the fact that the automatic object recognition using 3DI data has been an area of research for quite a few years with limited progress.

  Currently, the data acquisition time is much less than the time needed for post-processing. In addition, highly trained and skilled personnel are needed to use the software. Automation would not only reduce post-processing time but also reduce the dependence of the final results on the operator's skills.

- *Error characterization and how it affects measurements made in the software*

  For continued and expanded growth/use of 3DI systems, confidence in the measurements and deliverables based on these measurements is needed. The confidence in the measurements from these systems is addressed by the increased accuracy of the instruments. The confidence in the deliverables from 3D data can be addressed by the ability to quantify the error of the deliverables. This requires prior knowledge of the instrument's error and other sources of errors (e.g., registration) and the ability to propagate these errors to the end deliverable.

In the simplest case, the error in the distance between two points is on the same order as that of the instrument error. However, with 3DI systems, thousands of points are collected on an object. Therefore, the error of the distance between two walls (modeled as planes) benefits from the thousands of points used to define the planes and the error would be $\sqrt{N}$ (where N is the number of points) less than the error of the instrument.

- *Detailed as-built model*

Fitting an as-designed model to a point cloud is usually the first step in creating an as-built model. For example, the width of a rectangular room is determined as the distance between two parallel planes which represent the models of two opposite walls. In this example, the deviation between the as-built and the as-designed models could be defined as the difference between the as-built and as-designed room widths (see Figure 65 a and c). The as-built room width could be derived from the distance between two parallel planes fitted to the point cloud. In general, two walls are never constructed truly parallel. Therefore in this simplified modeling approach, possible local deviations of the actual room width are ignored.

The large number of points acquired by 3D imaging systems allows more detailed modeling which takes into account local deviations from the as-designed model. For example, the as-built walls may be modeled by higher order surfaces (e.g., NURBs) in which case the local deviations from the designed room width could be evaluated. In order to take full advantage of the acquired large datasets, the functionality of the post-processing software needs to be extended beyond simple geometrical modeling (e.g., NURB surfaces instead of planes or elliptic cylinders instead of circular cylinders). In addition, the ability to evaluate and display a map of local deviations of one surface from another is needed (see Figure 65d). Currently, most of the software packages display only a map of the distances of the data points from the as-designed or the as-built surfaces which may not allow for tracking local deviations between the as-built and as-designed models (especially when the deviations between both surfaces are comparable to or smaller than the instrument error).

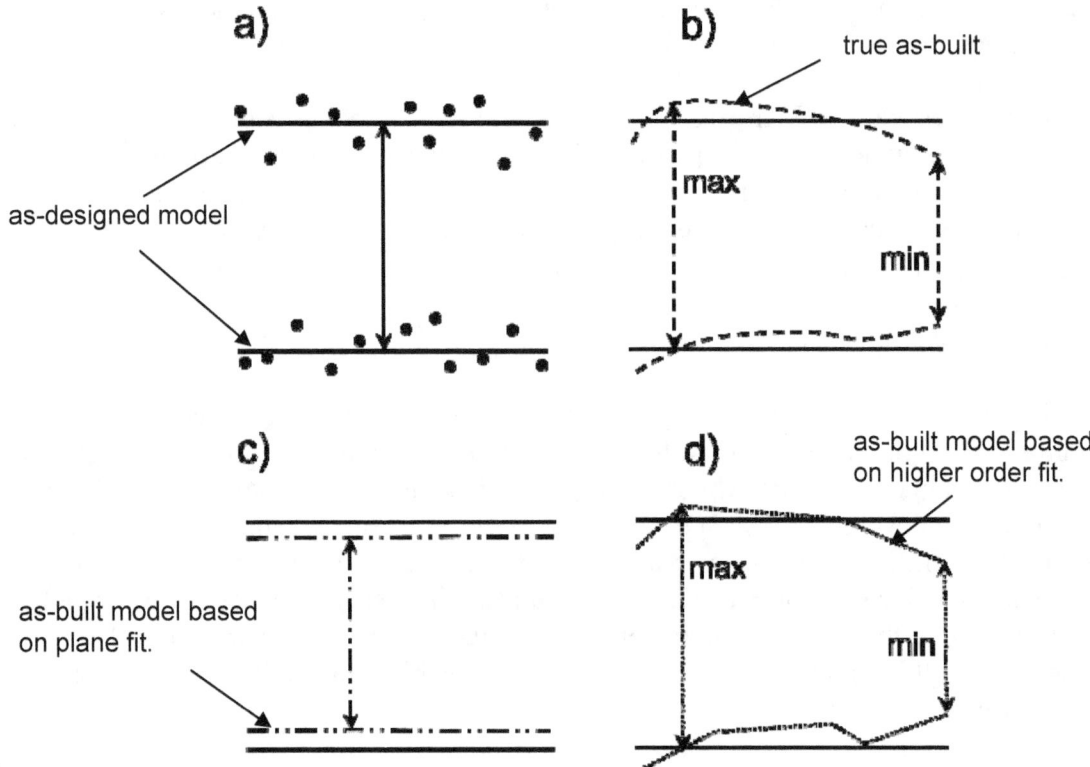

Figure 65. Schematic diagram of fitting a model to a point cloud: a) acquired point cloud registered to the as-designed model (represented here by a pair of parallel planes); b) true (i.e. unknown) as-built model overlaid on the as-designed model; c) as-built model based on fitting two parallel planes to the acquired point cloud (the distance between the fitted planes is different from the as-designed distance); d) a higher order fitted as-built model enables checking of local deviations between the as-built and the as-designed model (i.e., determine form error).

- *Parallelization of software and 64-bit architecture*

  As the technology advances and the capability of the 3DI systems to collect even more data increases, it is crucial that the software be able to efficiently process all these data. To increase the computational power of the computer, the current trend is to use multiple processors instead of using a single faster processor. Therefore, in order to increase the processing speed and improve efficiency, the software needs to be parallelized.

  A few software packages are written to take advantage of the 64-bit architecture. It is expected that this trend will be followed by the other software vendors.

- *Better registration workflow*

  Better registration workflows can be developed to improve work efficiency. Methods to improve the registration process include:

- The use of "smart targets", e.g. targets that can broadcast their ID and location to the 3DI system.
- Good field practices (e.g., practices to eliminate errors in identifying and documenting locations of targets)
- Software algorithms that can perform on-the-fly registration without targets

- *Compression of point cloud data*

  Because of the large amount of data generated by current 3DI systems and the current trend of faster data acquisition which means even more collected data, there is a need to develop an algorithm for context sensitive data compression which will allow the user to easily search the compressed databases. There is a trend for making point cloud data available in a BIM model. A primary purpose of creating a BIM model is the ability to obtain information about the facility throughout its lifecycle, i.e., from "cradle-to-grave" or from design to decommissioning. Therefore, having access to the raw data is important because the model created from the raw data is 1) a simplification and 2) future needs may require that a slightly different model be created.

- *Integration of point cloud data into BIM software*

  Originally, the point cloud software was developed as stand-alone applications where interoperability between this type of software and CAD software was accomplished through common file formats. This situation has evolved to the point where the software for processing point clouds are becoming integrated with CAD programs as plug-ins. As stated in the first bullet (Interoperability), it is anticipated that the use of BIM will increase, and a similar progression will be observed for the point processing software and BIM software. As a case in point, GSA (General Services Administration) recently worked with BIM-authoring vendors to ensure that their software was capable of exporting the data in a neutral data format, IFC (Industry Foundation Classes).

- *Point cloud simulator*

  A point cloud simulator that incorporates a 3DI system's characteristics (e.g., measurement uncertainty, beam width, beam divergence angles), environmental conditions, and surface characteristics (e.g., reflectivity, roughness) enables:
  - test measurement strategies (e.g., determine optimal instrument location and instrument settings)
  - better understanding of instrument errors
  - data validation

  Simulated data can also be used for algorithm testing.

- *Standard data sets for verification and algorithm testing*

  Standard data sets could be used to test and compare performance of different packages in an objective way. Publicly available benchmarks would provide users an objective means of selecting software suitable for their particular needs. Standard data sets would also stimulate further research and development of new algorithms.

Appendix A: Initial Survey of Active 3D Imaging Technology

Table A. 1. Commercial 3D Imaging Sensor Producers.

Commercial Source	Source Address	Source Contact Info.	Source Email & website	Types of Products	Products/Devices	Applications
CSEM, Swiss Center for Electronics and Microtechnology	Technoparkstrasse 1, CH-8005 Zurich, Switzerland	Nicolas Blanc +41 44 497 14 47	nicolas.BLANC@csem.ch www.csem.ch	FPA with custom CMOS/CCD AM-CW demodulation	SR3000 SR4000(see MESA Imaging)	Industrial robotics & automation Security inspection Consumer sub-cm applications
PMDTechnologies GmbH	Am Eichenhang 50, D-57076 Siegen, Germany	Robert Lange W:+49 271 238 538 815 Dr. Buxbaum W: +49 271 238 538 802	r.lange@PMDTec.com b.buxbaum@PMDTec.com www.PMDTec.com	AM-CW FPA CMOS compact 3D Imagers	PMD[vision] 03 PMD[vision] S3 PMD[vision] CamCube 3.0	Industrial robotics & automation Security inspection Consumer sub-cm applications
3DV Systems Ltd. (now part of Microsoft)	2nd Carmel St. Industrial Park Bldg.1 P.O. Box 249 Yokneam Il 20692	Giora Yahav, Ph.D. W:+972 49599599	giora@3dvsystems.com info@3dvsystems.com www.3dvsystems.com	Real-time TOF 3D cameras	Compact cameras Zcam	Gaming Web conferencing Robotics Automotive safety
Canesta, Inc.	440 N. Wolfe Rd Suite 101 Sunnyvale, CA 94085	Tony Zuccarino W:408 524 1430	tonyz@canesta.com www.canesta.com	Real-time AM-CW FPA CMOS compact 3D Imagers	Canesta Vision camera modules	Industrial robotics & automation Security inspection, Consumer sub-cm applications
LightTime, LLC	951 Old County Road Suite 299 Belmont, CA 94002	Robert Potenza W:510 217 8069	Potenza@LightTime.com www.LightTime.com	MEMS based scanning 3D imager	MEMScan 3D imager	Industrial robotics & automation Security Autonomous navigation Surveying

Commercial Source	Source Address	Source Contact Info.	Source Email & website	Types of Products	Products/Devices	Applications
MESA Imaging AG Acroname Inc	Technoparkstrasse 1 8005 Zurich, CH Boulder, CO	Thierry Oggier W:+41 44 508 1803 Suzanne Kaufmann W:	www.mesa-imaging.ch Suzanne@acroname.com	FPA with custom CMOS/CCD AM-CW demodulation	SR3000 SR4000	Industrial robotics & automation Security inspection Consumer sub-cm applications
Advanced Scientific Concepts Inc.	135 East Ortega St. Santa Barbara, CA 93101	Roger Stettner W:805 966 3331	rstettner@asc3d.com www.asc3d.com	64x64 and 128x128 APD Flash 3D imager InGaAs and Si PIN	Portable, miniature, underwater, super-high res.	Autonomous navigation Security Obscurant penetration Underwater inspect.
Z+F USA, Inc.	700 Old Pond Road Suite 606 Bridgeville, PA 15017	Eric De Jans W:412 257 8575	edejans@aol.com info@zf-usa.com www.zf-usa.com	AM phase based TOF High-res. Scanning 3D imager	Imager 5006i Imager 5010	Industrial sites Building documentation Infrastructure doc. Cultural heritage Forensics
Z+F GmbH	Simoniusstrasse 22 D-88239 Wangen im Allgau, DE	Christoph Frohlich +49(0)752293080	cf@zofre.de www.zofre.de	AM phase based TOF High-res Scanning 3D imager	Imager 5006i Imager 5010	Industrial sites Building documentation Infrastructure doc.
Autonosys Inc.		Robert Bruce W:613 482 6569	inquiries@autonosys.com www.autonosys.com	AM phase based TOF High-res. Scanning 3D imager	LVC0702 uses z+f rangefinder	Vehicle collision avoidance
Riegl USA, Inc.	7035 Grand National Dr. Suite 100 Orlando, Florida 32819	Ted Knaak CC: Sue Martin 407 248 9927	tknaak@rieglusa.com CC:smartin@rieglusa.com www.rieglusa.com	Pulse time-of-flight scanning 3D imagers	LMS-Q120i and ii, LMS-Q20, LMS-Q500, LMS-z210ii-S, VQ-180, VQ-250, VZ-400	Industrial sites Mobile applications
Riegl Laser Measurement Systems GmbH	Riedenburgstrasse 48 A-3580 Horn Austria	Michael Mayer International Sales +4329824212	mmayer@riegl.co.at www.riegl.com	Pulse time-of-flight scanning 3D imagers	LMS-Q120i and ii, LMS-Q20, LMS-Q500, LMS-z210ii-S, VQ-180, VQ-250, VZ-400	Industrial sites Mobile applications

Commercial Source	Source Address	Source Contact Info.	Source Email & website	Types of Products	Products/Devices	Applications
Neptec	302 Legget Drive Kanata, Ontario Canada K2K 1Y5	Iain Cristie W:613 599 7602	ichristie@neptec.com www.neptec.com	Triangulation and TOF based 3D imaging	LMS-SR LMS-MR	Industrial metrology systems
3rdTech, Inc.	2500 Meridian Parkway, Suite 150, Durham, NC 27713	Doug Schiff - VP Marketing W:919 361 2148	info@3rdTech.com dbs@3rdTech.com www.3rdTech.com	AM phase based TOF Scanning 3D imager	DeltaSphere -3000	Industrial sites Building documentation Infrastructure doc.
FARO Technologies	250 Technology Park Lake Mary, FL 32746	Robert Bridges 407 333 9911	robert.bridges@FARO.com www.FARO.com	AM phase based TOF Scanning 3D imagers	FARO Photon 120/20, FARO Laser Scanner LS880, Focus 3D	Infrastructure doc. Reverse engr. Inspection Heritage doc.
Optech Inc.	300 Interchange Way, Vaughan, Ontario Canada, L4K 5Z8	Wayne Szameitat sales Paul LaRocque 905 660 0808	waynes@optech.ca PaulL@optech.ca www.optech.ca	Pulse time-of-flight scanning 3D imager	ILRIS-3D ILRIS-HD ILRIS-LR	Industrial sites Building documentation Infrastructure doc. Surveying
Optech International Inc.	7225 Stennis Airport Dr. Suite 400 Kiln, MS 39556	Dr. Grady Tuell 228 252 1004	gradyt@optechint.com www.optechint.com	Pulse time-of-flight scanning 3D imager	ILRIS-3D ILRIS-HD ILRIS-LR	Industrial sites building documentation Infrastructure doc. Surveying
Leica Geosystems	HDS Engineering Solutions Eastern USA & Canada	Bruce Bowditch cell:616 510 1211	bruce.bowditch@lgshds.com www.leica-geosystems.us	Pulse time-of-flight scanning 3D imagers AM phase based TOF Scanning	Leica ScanStation C10 Leica HDS6200	Industrial sites Building documentation Infrastructure doc. Surveying
Leica Geosystems AG	Heinrich Wild Strasse CH-9435 Heerbrugg St. Gallen, Switzerland	Walter Schwyter CFO +41 71 727 3131	www.leica-geosystems .com	Pulse time-of-flight scanning 3D imagers AM phase based TOF Scanning 3D imagers	Leica ScanStation C10 Leica HDS6200	Industrial sites Building documentation Infrastructure doc. Surveying

Commercial Source	Source Address	Source Contact Info.	Source Email & website	Types of Products	Products/Devices	Applications
Maptek International Office in USA	165 S. Union Blvd. Suite 888 Lakewood, CO 80228	John Dolan 303 763 4919	john.dolan@maptek.com www.maptek.com	Pulse time-of-flight scanning 3D imager	I-Site 4400LR I-Site 4400CR I-Site 8800	Mining & industrial sites Building & infrastructure doc.
Measurement Devices US MDL	17555 Groeschke Rd Houston, TX 77084	Roger Bryant 281 646 0050	RBryant60@Gmail sales@mdl-laser.com www.mdl.co.uk	Pulse time-of-flight scanning 3D imager	QuarrymanPro C-ALS Cavity Scanner	Mine and quarry profiling Surveying of abandoned mines and cavities
Spatial Integrated Systems, Inc.	9055 Comprint Ct. Suite 220 Gaithersburg, MD 20877	Greg Walker & Joanna King 301 610 7965	greg.walker@sisinc.org joanna.king@sisinc.org www.sisinc.org	AM phase based TOF Scanning 3D imager	3DIS Model 1500	Building & infrastructure doc. Reverse engr. Inspection Forensics
Topcon Europe B.V.	Essebaan 11, 2908 LJ P.O. Box 145, 2900 AC Capelle a/d Ijssel, Netherlands	contact information +31 10 4585077	www.topcon-positioning.eu	Pulse time-of-flight scanning 3D imager	GLS-1500 Laser Scanner	Industrial sites Building documentation Infrastructure doc.
Topcon Positioning Systems	7400 National Dr. Livermore, CA 94550	Bryan Given 925 2453718	bgiven@topcon.com www.topconpositioning.com	Pulse time-of-flight scanning 3D imagers	GLS-1500 Laser Scanner	Industrial sites Building documentation Infrastructure doc.
Trimble Corp. Navigation Headquarter	935 Stewart Drive Sunnyvale, CA 94085	Tim Lemmon-plant products Bryan Williams-spatial imaging	Tim_Lemmon@trimble.com Bryan_Williams@trimble.com www.trimble.com	AM phase shift & pulse time-of-flight scanning 3D imagers	Trimble GX, Trimble FX, Trimble VX-Spatial, Trimble CX, Callidus 3D scanner	Surveying Construction Plant 3D mapping and documentation
Trimble GeoSpatial	10355 Westmoor Dr. Suite 100, Westminster, CO 80021	Omar Soubra 720 587 4517	omar_soubra@trimble.com www.trimble.com/geospatial	Pulse time-of-flight scanning 3D imager	Trimble Harrier MX8 Mobile	Aerial & mobile terrain modeling hardware and software

Commercial Source	Source Address	Source Contact Info.	Source Email & website	Types of Products	Products/Devices	Applications
Velodyne	345 Digital Drive Morgan Hill, CA 95037	Michael Dunbar 408 465 2859	mdunbar@velodyne.com www.velodyne.com/lidar	Pulse time-of-flight scanning 3D imager	HDL-64E HDL-32E	Unmanned vehicle Autonomous nav. Mapping/surveying
General Dynamics Robotic Systems	1231 Tech Court Westminster, MD 21157	410 876 9200 Barbara Lindauer VP Business Dev	blindauer@gdrs.com www.gdrs.com	Pulse time-of-flight scanning 3D imager		Unmanned systems Autonomous nav. Security
NextEngine, Inc.	401 Wilshire Blvd., Ninth Floor Santa Monica, CA 90401	Sarah Black 301 883 1888 Fax:310 883 1860	sarahb@nextengine.com www.nextengine.com	Laser Triangulation	Model 2020i	CAD part modeling CAM
Nvision, Inc.	440 Wrangler Drive Suite 200 Coppell, TX 75019	972 393 8000	sales@nvision3d.com www.nvision3d.com	High accuracy laser scanning Approach	MAXOS MobileScan 3D HandHeld Scanner	Reverse engr. Inspection
Nikon Metrology, Inc. (formerly Metris USA)	12701 Grand River Brighton, MI 48116	Anthony Scirpo 203 720 0010	tony.scirpo@nikonmetrology.com www.nikonmetrology.com	FM Coherent Laser Radar	MV224/260	Large Volume Metrology
H.N. Burns Engineering Corp.	3275 Progress Dr., Suite A Orlando, FL 32826	H.N."BUCK" Burns 407 273 3770	buck@hnbec.com	Pulse time-of-flight scanning 3D imager		Aerial & ground vehicle terrain surveying & security
Ibeo Automobile Sensor GmbH	Merkurrong 20 22143 Hamburg Germany	Mario Brumm +49 40 298 676 33	mario.brumm@ibeo-as.com www.ibeo-as.com	Pulse time-of-flight scanning 3D imager	Alasca XT scanner IBEO LUX	Automotive Safety & Security
Sick, Inc.	6900 West 110th St. Minneapolis, MN 55438	Stacy Kelly 1 800 325 7425	stacy.kelly@sick.com www.sickusa.com	Pulse time-of-flight scanning 3D imager	LMS 100 and others LD-MRS	Industrial automation safety & security
EADS Deutschland GmbH	88039 Friedrichshafen, Germany	Ingo Schwartz +49 7545 8 2871	ingo.schwatz@eads.com www.eads.net	Pulse time-of-flight scanning 3D imager	Hellas-Warning Hellas-Awareness	Aerial/Helicopter collision avoidance

Commercial Source	Source Address	Source Contact Info.	Source Email & website	Types of Products	Products/Devices	Applications
Dimensional Photonics International, Inc.	187 Ballardvale St. Wilmington, MA 01887	Lyle Shirley 978 988 8825	info@dpi-3d.com www.dimensionalphotonics.com also visit www.faro.com	Accordion Fringe Interferometry	AFI 5000P see FARO AFI MICRO	High res. reverse engr. Inspection Robotic applications Dentistry
MD3D Limited	Oakmead House Pangbourne Rd Upper Basildon RG88LN UK	Mike Davies +44 1491 671800	sales@md3d.uk.com www.md3d.uk.com	Phase based TOF High-res. Scanning digital fringe projection	Surphaser 25 HSX Kolibri	Reverse Engr. Inspection metrology
Basis Software, Inc.	2811 152nd Ave NE Redmond, WA 98052	Peter Petrov 425 8619390	petrov@basissoftware.com info@surphaser.com www.surphaser.com	Phase based TOF High-res. Scanning digital fringe projection	Surphaser 25 HSX Kolibri	Reverse Engr. Inspection Metrology
Laser Design Inc.	9401 James Ave. South Suite 132 Minneapolis , MN 55431	Giles Gaskell at GKS 734 5829600 LDI - 952 8849648	ggaskell@gks.com www.gks.com sales@laserdesign.com www.laserdesign.com	Laser line scanning	Surveyor WS-series, DM-series, DS-series	Reverse Engr. Inspection Metrology
Direct dimensions	10310 S. Dolfield Rd. Owings Mills, MD 21117	Michael Raphael 410 998 0880 X103	mraphael@dirdim.com www.dirdim.com	Phase based TOF High-res. Scanning	Surphaser 25 HSX and other scanners	Reverse Engr. Inspection Metrology

Table A. 2. 3D Imaging Research Systems.

Source Name	Source Address	Source Contact Info.	Source Email & website	3D Imaging Technology	Products/Devices	Applications
Lockheed Martin Coherent Technologies	135 South Taylor Ave. Louisville, CO 80027	Duane Smith Brian Redman Philip Gatt 303 604 2000	duane.d.smith@lmco.com brian.c.redman@lmco.com philip.gatt@lmco.com	Coherent FM-CW LADAR FPA holography based LADAR	Prototypes and products for military	Primarily defense and security
ARL Sensors Directorate, Army Research Lab	Adelphi, MD 20783	Barry Stann 301 394 3141	stann@arl.army.mil	FPA Chirped AM (FM/CW) LADAR MEMS LADAR	Prototypes and products for military	Primarily defense and security
NIST	325 Broadway Boulder, CO 80305	W.C. Swann and N.R. Newbury 303 497 4227	nnewbury@boulder.nist.gov	Femtosecond freq. comb LIDAR	research prototype	Long distance high-res. range imaging
Sandia National Laboratories	P.O. Box 5800 Albuquerque, New Mexico 87185	John V. Sandusky 505 845 0132	jvsandu@sandia.gov	non-scanning intensified CCD FPA imagers - multi-freq. AM-CW	SRI Flash Quad others LDRI	Real-time high-res. range imaging
BAE Systems Advanced Systems and Technology	P.O.Box 868, MER15-2415 Nashua, New Hampshire 03061	George Dippel 603 885 6638	george.dippel@baesystems.com	non-scanning CCD FPA imagers – using coherent laser approaches	Prototypes for military	Sub-mm range imaging
Night Vision and Electronic Sensors Dir.	AMSRD CER NV SPP 10221 Burbeck Road Ft. Belvoir, Virginia 22060	J. Andrew Hutchinson 703 704 3249	andy.hutchinson@nvl.army.mil	Multi-pulse LADARs	Prototypes for military	Range imaging and foliage penetration
MIT Lincoln Lab.	244 Wood Street Lexington, MA 02420	Dr. Richard Heinrichs 781 981 7945 Dr. Richard Marino 781 981 4011	hieinrichs@ll.mit.edu marino@ll.mit.edu	Geiger mode APD array LADAR other coherent laser approaches	Prototypes for military	Range imaging and foliage penetration
Lumen Labs. Inc.	103 Terrace Hall Ave. Burlington, MA 01803	Robert Dillon 781 273 5995	www.lumenlabs.com			

Source Name	Source Address	Source Contact Info.	Source Email & website	3D Imaging Technology	Products/Devices	Applications
Sensors Unlimited	3490 Route 1 Building 12 Princeton, NJ 08540	Robert Struthers 609 524 0227	robert.struthers@goodrich.com www.oss.goodrich.com	FPA Flash LADAR APD	APD arrays	Range imaging for defense and security
DRS Technologies Infrared Technologies, LP	13544 N. Central Expressway Dallas, Texas 75243	Jeffrey D. Beck 972 560 5988	jdbeck@drs-irtech.com www.drs.com	active pulsed and passive FPA APD arrays	APD arrays	Range imaging for defense and security
Raytheon Vision Systems	75 Coromar Dr. Bldg. 2, Mail Station 8 Goleta, California 93117	Michael D. Jack 805 562 2395	mdjack@raytheon.com or patrotta@west.raytheon.com	FPA Flash LADAR	Mercury Cadmium Telluride Flash Array system 256x256	Range imaging for defense and security
Intevac Photonics Tech. Div.	3560 Bassett St. Santa Clara, CA 95054	Verle Aebi - Pres. Steve Campano Market 408 986 9888	vaebi@intevac.com scampano@intevac.com	Miniature Photocathode detectors	Range gated array imagers	Range imaging for defense and security
Lockheed Martin Advanced Tech. Directorate	P.O. Box 650003 M/S: PT-88 Dallas, Texas 75265	Bruno Evans 972 603 7945	bruno.evans@lmco.com	Small scanned pulse laser TOF LADAR	InGaAs or Silicon detectors breadboard systems	Range imaging for defense and security obscurant penetration
iRobot	63 South Ave. MS-112 Burlington, MA 01803	Edison Hudson 781 418 3409 or 781 428 3351	ehudson@irobot.com www.irobot.com	Small FPA Flash LADAR InGaAs APD detector	Robo-i breadboard system	Range imaging for small robot navigation needs
Northrop Grumman Space Systems Division	1100 West Hollyvale St. P.O. Box 296 Azusa, CA 91702	Raj Shori 626 812 2990	raj.shori@ngc.com			

Appendix B: 2010 Technical Survey of Active 3D Imaging

Products and Prototypes

Manufacturer's Name: Army Research Lab - Research Prototype	
Product Name and Model Number: None – MEMS-Scanned LADAR Sensor Prototype	
System Architecture: Uses pulsed laser (2 ns to 3 ns pulse) to determine range to a pixel and a two-axis MEMS mirror to establish the angular direction to a pixel.	
Applications: Low-cost, compact, low-power LADAR imager primarily for small unmanned ground vehicles for navigation, obstacle/collision avoidance, and target detection and identification. Possible commercial uses for small autonomous robots in household and industrial applications.	
Year Introduced: 2009	
Wavelength of illumination source (nm)	1550 nm
Illumination power (mW)	0.4 W average, 1 kW peak
Laser safety classification	TBD
Beam diameter at exit (mm)	0.5 mm
Beam divergence (mrad)	1 mrad
Min./Max. range (m)	1m / 20m
Range uncertainty (mm) (at specified range, reflectivity, and number of measurements)	12 mm rms at 10 m, .5 reflectivity, 32 k samples
Range resolution (depth) [1] (mm)	Currently 420 mm (based on FWHM pulse width)
Can system provide range profile (multiple returns) for the same range column (Y/N)	Range and amplitude data are collected for every .1 m in range. Currently up to 3 targets in a single range profile are displayed.
Sensor field of view, horizontal (degrees)	40°, 60° demonstrated
Sensor field of view, vertical (degrees)	30°
Pixel or date acquisition rate (pxl/s) (if scanner used)	(200 to 400) kHz
Frame rate (if FPA) (frames/s)	(6 to 12) Hz for a 256x128 pixel image
Array size (if FPA) (pixels)	N/A
Angular uncertainty (degrees)	TBD
Angular resolution (degrees)	0.156°, horizontal
Color imagery availability/registration (Y/N)	N
Rated operating conditions (such as temperature, ambient brightness, etc.)	TBD
Limiting operating conditions	
Data and communication interfaces	Ethernet
Power supply voltage (V) and consumption (W)	5 V, 30 W
Overall size of unit (w x d x h) (mm)	(95x200x50) mm for prototype sensor modules
Weight of complete sensor (kg)	1 kg
Retail cost (U.S. $)	$12.6K for components in small quantities
Lead time to delivery (weeks)	Experimental LADAR
What processing software is provided (besides control software)	TBD

Manufacturer's Name: Advanced Scientific Concepts, Inc.	
Product Name and Model Number: Portable 3D Flash LIDAR Evaluation Kit	
System Architecture: InGaAs APD detector array with CMOS Read Out IC; 3D Flash LIDAR camera (≈ 5 ns laser pulse) – with cooling options (air or water)	
Applications: Autonomous navigation, security, obscurant penetration, inspection, mapping, aviation situational awareness, target ID and tracking,	
Year Introduced: 2006	
Wavelength of illumination source (nm)	1570 nm, 1060 nm optional
Illumination power (mW)	(2.5 to 5) mJ/pulse
Laser safety classification	Eye-safe
Beam diameter at exit (mm)	4.25 mm square
Beam divergence (mrad)	Depends on diffuser (1.5, 3, 8.6 or 45 degree FOV)
Min./Max. range (m)	70 m, 300 m, 600 m, 1100 m depends on selected FOV
Range uncertainty (mm) (at specified range, reflectivity, and number of measurements)	(20 to 30) mm
Range resolution (depth) [1] (mm)	18 cm
Can system provide range profile (multiple returns) for the same range column (Y/N)	Y: Because all the pixels (entire scene) is captured < 10nS with each laser pulse, it may not be necessary
Sensor field of view, horizontal (degrees)	1.5°, 3°, 8.6° or 45°
Sensor field of view, vertical (degrees)	Same as above
Pixel or date acquisition rate (pxl/s) (if scanner used)	Nominal = 327 000 pixels per second (non-scanning, staring camera, solid state)
Frame rate (if FPA) (frames/s)	(1 to 20) Hz typical, 30 Hz possible in burst mode
Array size (if FPA) (pixels)	128x128
Angular uncertainty (degrees)	Small, almost immeasurable
Angular resolution (degrees)	0.012°, 0.023°, 0.070° or 0.35°
Color imagery availability/registration (Y/N)	Y with co-aligned IR or visible camera
Rated operating conditions (such as temperature, ambient brightness, etc.)	0 °C to 50 °C
Limiting operating conditions	
Data and communication interfaces	Camera Link
Power supply voltage (V) and consumption (W)	2 x 120 W 120 VAC for camera and laser
Overall size of unit (w x d x h) (mm)	(6.5 x 12 x 10) cm
Weight of complete sensor (kg)	6.5 kg
Retail cost (U.S. $)	≈US$ 285 000
Lead time to delivery (weeks)	16 weeks ARO
	Requires external PC notebook for processing data
What processing software is provided (besides control software)	Flash3D Licensed Software, can be used with standard 3D computer graphics tools

Manufacturer's Name: Advanced Scientific Concepts, Inc.		
Product Name and Model Number: TigerEye 3D Flash LIDAR Camera Kit		
System Architecture: InGaAs APD detector array with CMOS Read Out IC: 3D flash LIDAR (≈5 ns laser pulse) – with active or passive cooling		
Applications: Autonomous navigation (Space AR&D, UAV, UGV), security, obscurant penetration, inspection, mapping, aviation (helicopter landing, target ID and tracking, etc.), situational awareness		
Year Introduced: 2010		
Wavelength of illumination source	(nm)	1570 nm
Illumination power	(mW)	(2.5 to 8) mJ/pulse
Laser safety classification		Class I Eye-safe
Beam diameter at exit	(mm)	3 mm square
Beam divergence	(mrad)	Depends on diffuser (i.e. 45, 8.6, 3 degree FOV)
Min./Max. range	(m)	70m, 450m, 1,250m depending on selected FOV
Range uncertainty (at specified range, reflectivity, and number of measurements)	(mm)	(20 to 30) mm
Range resolution (depth) [1]	(mm)	≈3 cm to 5 cm
Can system provide range profile (multiple returns) for the same range column	(Y/N)	Y: Because all the pixels (entire scene) is captured < 10nS with each laser pulse, it may not be necessary
Sensor field of view, horizontal	(degrees)	45 (17mm), 8.6 (85 mm), 3 (250 mm) degrees
Sensor field of view, vertical	(degrees)	Same as above, 22 x 45 degree FOV optional
Pixel or date acquisition rate (if scanner used)	(pxl/s)	Nominal = 327,000 pixels per second (non-scanning, staring camera, solid state)
Frame rate (if FPA)	(frames/s)	(1 to 20) Hz typical; 30 Hz optional
Array size (if FPA)	(pixels)	128x128 (128x64 for 45x22 FOV unit)
Angular uncertainty	(degrees)	So small, almost immeasurable
Angular resolution	(degrees)	0.35 (45 FOV); 0.67 (8.6 FOV); 0.0234 (3 FOV)
Color imagery availability/registration	(Y/N)	Y (with optional, co-aligned 2D or IR cameras)
Rated operating conditions (such as temperature, ambient brightness, etc.)		0 °C to 50 °C
Limiting operating conditions		NA; Space qualified for Low Earth Orbit
Data and communication interfaces		Ethernet/Power interface
Power supply voltage (V) and consumption (W)		24 VDC (+/- 4V), (28 to 45) W dependent upon cooling
Overall size of unit (w x d x h)	(mm)	(110 x 112 x 121) mm
Weight of complete sensor	(kg)	<2 kg
Retail cost	(U.S. $)	≈$150 000
Lead time to delivery	(weeks)	14 weeks ARO
		All processing done "on camera"
What processing software is provided (besides control software)		ASC 2D and 3D viewing software (TigerView) and Flash3D command & control software

Manufacturer's Name: Basis Software		
Product Name and Model Number: Surphaser 25HSX (multiple configurations)		
System Architecture: Multiple-frequency AMCW TOF, plannar-mirror/pan scanning mechanism		
Applications: Reverse engineering, inspection, metrology		
Year Introduced: 2006		
Wavelength of illumination source	(nm)	685 nm
Illumination power	(mW)	15 mW
Laser safety classification		3R
Beam diameter at exit	(mm)	2.3 mm
Beam divergence	(mrad)	0.1 mrad
Min./Max. range	(m)	2m/46m
Range uncertainty (at specified range, reflectivity, and number of measurements)	(mm)	0.33 mm at 10m (very low noise)
Range resolution (depth) [1]	(mm)	0.001 mm under best conditions
Can system provide range profile (multiple returns) for the same range column	(Y/N)	N
Sensor field of view, horizontal	(degrees)	(0 to 360)°
Sensor field of view, vertical	(degrees)	270°
Pixel or date acquisition rate (if scanner used)	(pxl/s)	214,000 to 1,000,000 samples/s
Frame rate (if FPA)	(frames/s)	N/A
Array size (if FPA)	(pixels)	N/A
Angular uncertainty	(degrees)	0.01°
Angular resolution	(degrees)	8 arc sec (0.0022°)
Color imagery availability/registration	(Y/N)	
Rated operating conditions (such as temperature, ambient brightness, etc.)		0.5 °C to 45 °C, primarily for indoor light conditions
Limiting operating conditions		
Data and communication interfaces		USB 2.0
Power supply voltage (V) and consumption (W)		(19 to 24) V (65 W peak)
Overall size of unit (w x d x h)	(mm)	(285x170x480) mm
Weight of complete sensor	(kg)	11 kg
Retail cost	(U.S. $)	$95K
Lead time to delivery	(weeks)	4-12 weeks
What processing software is provided (besides control software)		Typically used with commercially available SW

Manufacturer's Name: Boeing	
Product Name and Model Number: Small SWAP LADAR UAV Mapping Payload	
System Architecture: Scanning 32x32 Photon Counting Array Mapper.	
Applications: Mapping and 3D imaging where small size, weight and power are required.	
Year Introduced: 2010	
Wavelength of illumination source (nm)	1064
Illumination power (mW)	2000 mW
Laser safety classification	IIIb
Beam diameter at exit (mm)	3 mm
Beam divergence (mrad)	23
Min./Max. range (m)	100 m / 10,000 m
Range uncertainty (mm) (at specified range, reflectivity, and number of measurements)	
Range resolution (depth) [1] (mm)	150 mm (timing clocks for Geiger latches are at 2 GHz, jitter in clocks contribute to depth spread)
Can system provide range profile (multiple returns) for the same range column (Y/N)	yes
Sensor field of view, horizontal (degrees)	20
Sensor field of view, vertical (degrees)	1.3
Pixel or date acquisition rate (pxl/s) (if scanner used)	8M
Frame rate (if FPA) (frames/s)	8000
Array size (if FPA) (pixels)	32x32
Angular uncertainty (degrees)	
Angular resolution (degrees)	0.04
Color imagery availability/registration (Y/N)	N
Rated operating conditions (such as temperature, ambient brightness, etc.)	
Limiting operating conditions	Design trades meet operational requirements
Data and communication interfaces	Camera Link, Ethernet
Power supply voltage (V) and consumption (W)	24 V, 325 W
Overall size of unit (w x d x h) (mm)	(152 x 558 x 152) mm
Weight of complete sensor (kg)	9
Retail cost (U.S. $)	<$1M (ROM)
Lead time to delivery (weeks)	12 (ROM)
What processing software is provided (besides control software)	none

Manufacturer's Name: Boeing Research & Technology		
Product Name and Model Number: Targetless Laser Locator System		
System Architecture: Laser Radar precision range & precision angle measurement. Two parts to system: Optical Head Assembly (OHA) containing scanner and laser radar engine; Electronic Support Chassis (ESU) containing embedded computer, electronics and power.		
Applications: Precision dimensional metrology for aerospace production and test.		
Year Introduced: 2008		
Wavelength of illumination source	(nm)	1550
Illumination power	(mW)	Eyesafe
Laser safety classification		
Beam diameter at exit	(mm)	
Beam divergence	(mrad)	
Min./Max. range	(m)	0.5 m/7.6 m
Range uncertainty (at specified range, reflectivity, and number of measurements)	(mm)	< +/- 0.025 mm to 7.6 m (2 sigma)
Range resolution (depth) [1]	(mm)	
Can system provide range profile (multiple returns) for the same range column	(Y/N)	Y
Sensor field of view, horizontal	(degrees)	Depends on orientation: +/- 175° Elevation, +/-20°
Sensor field of view, vertical	(degrees)	See above
Pixel or date acquisition rate (if scanner used)	(pxl/s)	Range update rate: 6000 Hz; angle step time approx. 1.5 ms
Frame rate (if FPA)	(frames/s)	
Array size (if FPA)	(pixels)	
Angular uncertainty	(degrees)	45 PPM (2 sigma)
Angular resolution	(degrees)	
Color imagery availability/registration	(Y/N)	N- B&W video camera image; NRK'SA Visualization SW
Rated operating conditions (such as temperature)		Typical industrial environment
Limiting operating conditions		
Data and communication interfaces		Ethernet 10/100/GE
Power supply voltage (V) and consumption (W)		110 V AC
Overall size of unit (w x d x h)	(mm)	OHA: (584x216x368) mm; ESU: (510x292x152) mm
Weight of complete sensor	(kg)	OHA: 25.9 kg, ESU: 17.2 kg
Retail cost	(U.S. $)	Not for external sale
What processing software is provided (besides control software)		Visualization SW

Manufacturer's Name: Bridger Photonics, Inc.		
Product Name and Model Number: SLM-L		
System Architecture: FMCW LADAR; system measures range profile along line-of-sight; includes laser, receiver electronics and post-processing hardware		
Applications: Industrial metrology, precision manufacturing, surface characterization, machine vision and 3D imaging and object recognition		
Year Introduced:2010		
Wavelength of illumination source	(nm)	1550
Illumination power	(mW)	<10 mW standard. Higher power available custom
Laser safety classification		1M
Beam diameter at exit	(mm)	0.01 fiber output, (25 to 50) mm with additional optics
Beam divergence	(mrad)	300 fiber output, 0.1 with additional optics
Min./Max. range	(m)	Range window ≈5 m for 1 kHz update rate, Range baseline >10 km
Range uncertainty (at specified range, reflectivity, and number of measurements)	(mm)	<0.005 mm, (1 m range, 4 % reflectivity, 1 measurement, 1 ms data acquisition), measured as a standard deviation of successive measurements of the same measurand under repeatable conditions
Range resolution (depth) [1]	(mm)	1.5 mm, calculated as FWHM of range peak
Can system provide range profile (multiple returns) for the same range column	(Y/N)	Y
Sensor field of view, horizontal	(degrees)	N/A
Sensor field of view, vertical	(degrees)	N/A
Pixel or data acquisition rate (if scanner used)	(pxl/s)	N/A
Frame rate (if FPA)	(frames/s)	N/A
Array size (if FPA)	(pixels)	N/A
Angular uncertainty	(degrees)	N/A
Angular resolution	(degrees)	N/A
Color imagery availability/registration	(Y/N)	N
Rated operating conditions (such as temperature, ambient brightness, etc.)		22 °C ± 8 °C to maintain accuracy
Limiting operating conditions		Low vibration environment required for standard system. Vibration compensation available custom.
Data and communication interfaces		Communication through control software only
Power supply voltage (V) and consumption (W)		110 V AC, 20W
Overall size of unit (w x d x h)	(mm)	100 mm H x 480 mm W x 480 mm D
Weight of complete sensor	(kg)	8 kg
Retail cost	(U.S. $)	Contact Us
Lead time to delivery	(weeks)	12-16
What processing software is provided (besides control software)		None

Manufacturer's Name: Bridger Photonics, Inc.		
Product Name and Model Number: SLM-M		
System Architecture: FMCW LADAR; system measures range profile along line-of-sight; includes laser, receiver electronics and post-processing hardware		
Applications: Industrial metrology, precision manufacturing, surface characterization, machine vision and 3D imaging and object recognition		
Year Introduced:2010		
Wavelength of illumination source	(nm)	1550
Illumination power	(mW)	<10 mW standard. Higher power available custom
Laser safety classification		1M
Beam diameter at exit	(mm)	0.01 fiber output, 25-50 with additional optics
Beam divergence	(mrad)	300 fiber output, 0.1 with additional optics
Min./Max. range	(m)	Range window ≈100 m, Range baseline >10 km
Range uncertainty (at specified range, reflectivity, and number of measurements)	(mm)	< 2e-4 mm, (0.75 m range, 4 % reflectivity, 1 measurement, 60 ms data acquisition), measured as a standard deviation of successive measurements of the same measurand under repeatable conditions.
Range resolution (depth) [1]	(mm)	0.050 mm, calculated as FWHM of range peak (no windowing)
Can system provide range profile (multiple returns) for the same range column	(Y/N)	Y
Sensor field of view, horizontal	(degrees)	N/A
Sensor field of view, vertical	(degrees)	N/A
Pixel or data acquisition rate	(pxl/s)	N/A
Frame rate (if FPA)	(frames/s)	N/A
Array size (if FPA)	(pixels)	N/A
Angular uncertainty	(degrees)	N/A
Angular resolution	(degrees)	N/A
Color imagery availability/registration	(Y/N)	N
Rated operating conditions (such as temperature, ambient brightness, etc.)		22 °C ± 8 °C to maintain accuracy
Limiting operating conditions		Low vibration environment required for standard system. Vibration compensation available custom.
Data and communication interfaces		Communication through control software only
Power supply voltage (V) and consumption (W)		110 V AC, <150W
Overall size of unit (w x d x h)	(mm)	100 mm H x 480 mm W x 480 mm D
Weight of complete sensor	(kg)	8 kg
Retail cost	(U.S. $)	Contact Us
Lead time to delivery	(weeks)	12-16
What processing software is provided (besides control software)		None

Manufacturer's Name: EADS – Deutschland GmbH		
Product Name and Model Number: HELLAS-A (Awareness)		
System Architecture: 2D Scanning LADAR System with pulsed fibre Laser and APD detector; integral part of aircraft avionic. Classification of obstacles; display on MFD and/or HMS/D.		
Applications: Obstacle warning for military helicopters (contracted for NH90)		
Year Introduced: 2009		
Wavelength of illumination source (nm)	1550	
Illumination power (W)	15000	
Laser safety classification	Class 1	
Beam diameter at exit (mm)	48	
Beam divergence (mrad)	1.50	
Min./Max. range (m)	50 m/1200 m	
Range uncertainty (mm) (at specified range, reflectivity, and number of measurements)	± 600	
Range resolution (depth) [1] (mm)	600	
Can system provide range profile (multiple returns) for the same range column (Y/N)	N	
Sensor field of view, horizontal (degrees)	36 + 24 degrees line of sight steering	
Sensor field of view, vertical (degrees)	42°	
Pixel or date acquisition rate (pxl/s) (if scanner used)	Approx. 64000	
Frame rate (if FPA) (frames/s)	3	
Array size (if FPA) (pixels)	N/A	
Angular uncertainty (degrees)	± 0.16°	
Angular resolution (degrees)	0.33°	
Color imagery availability/registration (Y/N)	no	
Rated operating conditions (such as temperature, ambient brightness, etc.)	-40 °C + 50 °C	
Limiting operating conditions		
Data and communication interfaces	Arinc, Video, RS422, Milbus	
Power supply voltage (V) and consumption (W)	28 V, max 280 W, 180 W typical	
Overall size of unit (w x d x h) (mm)	320 x 300 x 320	
Weight of complete sensor (kg)	22.5 kg	
Retail cost (U.S. $)	Customer specific	
Lead time to delivery (weeks)		
What processing software is provided (besides control software)		

Manufacturer's Name: EADS – Deutschland GmbH	
Product Name and Model Number: HELLAS-W (Warning)	
System Architecture: 2D Scanning LADAR System with pulsed fibre Laser and APD detector. Classification and display of High Risk obstacles.	
Applications: Obstacle warning for helicopters (in operational use at German Federal Police) -Surveillance applications	
Year Introduced: 2002	
Wavelength of illumination source (nm)	1550
Illumination power (W)	4000
Laser safety classification	Class 1
Beam diameter at exit (mm)	60
Beam divergence (mrad)	1.50
Min./Max. range (m)	50 m/1200 m
Range uncertainty (mm) (at specified range, reflectivity, and number of measurements)	± 600
Range resolution (depth) [1] (mm)	600
Can system provide range profile (multiple returns) for the same range column (Y/N)	N
Sensor field of view, horizontal (degrees)	31.5°
Sensor field of view, vertical (degrees)	32°
Pixel or date acquisition rate (pxl/s) (if scanner used)	Approx. 55000
Frame rate (if FPA) (frames/s)	2
Array size (if FPA) (pixels)	N/A
Angular uncertainty (degrees)	± 0.16°
Angular resolution (degrees)	0.33° / 0.16°
Color imagery availability/registration (Y/N)	yes
Rated operating conditions (such as temperature, ambient brightness, etc.)	-25 °C + 55 °C
Limiting operating conditions	
Data and communication interfaces	Arinc, Video, Ethernet
Power supply voltage (V) and consumption (W)	28 V, max 250 W, 140 W typical
Overall size of unit (w x d x h) (mm)	320 x 450 x 320
Weight of complete sensor (kg)	26.4
Retail cost (U.S. $)	Customer specific
Lead time to delivery (weeks)	From stock
What processing software is provided (besides control software)	

Manufacturer's Name: FARO	
Product Name and Model Number: Laser Scanner Focus	
System Architecture: Phase TOF 3D imaging scanner (Ambiguity interval – 153.49 m) Stand-alone design	
Applications: 3D documentation of building construction, excavation volumes, façade and structural deformations, crime scenes, accident details, product geometry, factories, process plants.	
Year Introduced: 2010	
Wavelength of illumination source (nm)	905
Illumination power (mW)	20 mW
Laser safety classification	3R
Beam diameter at exit (mm)	3.8 mm, circular
Beam divergence (mrad)	0.16 mrad
Min./Max. range (m)	0.6 m to 120 m at 90 % reflectance 0.6 m to 20 m at 10 % reflectance
Range uncertainty (mm) (at specified range, reflectivity, and number of measurements)	± 2 mm at 10 m and 25 m, each at 90 % and 10 % reflectivity
Range resolution (depth) [1] (mm)	
Can system provide range profile (multiple returns) for the same range column (Y/N)	N
Sensor field of view, horizontal (degrees)	360°
Sensor field of view, vertical (degrees)	305°
Pixel or date acquisition rate (pxl/s) (if scanner used)	Variable: 122k / 244k / 488k / 976k pxl/s
Frame rate (if FPA) (frames/s)	N/A
Array size (if FPA) (pixels)	N/A
Angular uncertainty (degrees)	
Angular resolution (degrees)	0.009°
Color imagery availability/registration (Y/N)	Y (up to 70 megapixel)
Rated operating conditions (such as temperature, ambient brightness, etc.)	5 °C to 40 °C
Limiting operating conditions	
Data and communication interfaces	SD, SDHC, SDXC, 32 GB card included
Power supply voltage (V) and consumption (W)	19 V (external), 14.4 V (internal) – 40 W
Overall size of unit (w x d x h) (mm)	240 x 200 x 100
Weight of complete sensor (kg)	5.0 kg
Retail cost (U.S. $)	
Lead time to delivery (weeks)	
What processing software is provided (besides control software)	

Manufacturer's Name: Heliotis AG	
Product Name and Model Number: Optomoscope M2	
System Architecture: Fast 3D Microscopy using Parallel Optical low-Coherence Tomography (pOCT)	
Applications: Inspection of: micro-optical components, joints and adhesive bonds **Quality control of:** micromechanical pieces, injection molded parts	
Year Introduced:	
Wavelength of illumination source (nm)	Superluminescent Light Emitting Diode (SLED), 840
Illumination power (mW)	
Laser safety classification	Eye safe
Beam diameter at exit (mm)	
Beam divergence (mrad)	
Min./Max. range (m)	
Range uncertainty (mm) (at specified range, reflectivity, and number of measurements)	
Range resolution (depth) [1] (mm)	< 2 µm
Can system provide range profile (multiple returns) for the same range column (Y/N)	
Sensor field of view, horizontal (degrees)	From (3.8 x 3.2) mm to (0.2 x 0.2) mm depending on objective used: 2.5x, 10x, 40x, or 80x
Sensor field of view, vertical (degrees)	See above
Pixel or date acquisition rate (pxl/s) (if scanner used)	N/A
Frame rate (if FPA) (frames/s)	5000 2D slices per second, up to 5 3D fps
Array size (if FPA) (pixels)	
Angular uncertainty (degrees)	
Angular resolution (degrees)	From 30 µm to 2 µm depending on objective used
Color imagery availability/registration (Y/N)	
Rated operating conditions (such as temperature, ambient brightness, etc.)	
Limiting operating conditions	
Data and communication interfaces	USB 2.0
Power supply voltage (V) and consumption (W)	(110 to 240) VAC , 350 W (with PC)
Overall size of unit (w x d x h) (mm)	L 105 x W 124 x H 196 (head only)
Weight of complete sensor (kg)	2.5 kg (head only)
Retail cost (U.S. $)	
Lead time to delivery (weeks)	
What processing software is provided (besides control software)	

Manufacturer's Name: Heliotis AG	
Product Name and Model Number: M3-XL Optical Profiler	
System Architecture: Fast 3D profiler using Parallel Optical low-Coherence Tomography (pOCT)	
Applications: Inspection and quality control in microelectronics and miniaturized surfaces, micromechanical devices, and micro-optics. Tomographic and topological imaging of biological samples. 3D imaging for forensics and in security.	
Year Introduced: 2010	
Wavelength of illumination source (nm)	Superluminescent Light Emitting Diode (SLED), 800
Illumination power (mW)	8 mW
Laser safety classification	Eye safe
Beam diameter at exit (mm)	
Beam divergence (mrad)	
Min./Max. range (m)	
Range uncertainty (mm) (at specified range, reflectivity, and number of measurements)	
Range resolution (depth) [1] (mm)	1 µm (standard configuration, 20 nm optional)
Can system provide range profile (multiple returns) for the same range column (Y/N)	
Sensor field of view, horizontal (degrees)	(0.6 x 0.6) mm (standard configuration)
Sensor field of view, vertical (degrees)	See above
Pixel or date acquisition rate (pxl/s) (if scanner used)	N/A
Frame rate (if FPA) (frames/s)	Smart pixel sensor processing of up to 1 million 2D slices per second
Array size (if FPA) (pixels)	
Angular uncertainty (degrees)	
Angular resolution (degrees)	2 µm lateral resolution (standard configuration)
Color imagery availability/registration (Y/N)	Y Live view supports navigation on sample
Rated operating conditions (such as temperature, ambient brightness, etc.)	
Limiting operating conditions	
Data and communication interfaces	
Power supply voltage (V) and consumption (W)	
Overall size of unit (w x d x h) (mm)	
Weight of complete sensor (kg)	
Retail cost (U.S. $)	
Lead time to delivery (weeks)	
What processing software is provided (besides control software)	

Manufacturer's Name: Ibeo Automotive Systems GmbH	
Product Name and Model Number: Ibeo LUX Laserscanner family	
System Architecture: Laser pulse time-of-flight, multi layer scanning 3D imager	
Applications: Obstacle detection, identification, and tracking for automotive safety systems	
Year Introduced: 2009	
Wavelength of illumination source (nm)	905 nm (approx. 4.5 ns pulse)
Illumination power (mW)	Not provided
Laser safety classification	Class 1 (eye safe)
Beam diameter at exit (mm)	Not provided
Beam divergence (mrad)	0.08°
Min./Max. range (m)	0.3 to 50 m (@ 10 % remission) 200 m max.
Range uncertainty (mm) (at specified range, reflectivity, and number of measurements)	1sigma repeat accuracy 100 mm
Range resolution (depth) [1] (mm)	40 mm
Can system provide range profile (multiple returns) for the same range column (Y/N)	3 returns
Sensor field of view, horizontal (degrees)	110°
Sensor field of view, vertical (degrees)	3.2° or 6.4° (4 or 8 layer sensor)
Pixel or date acquisition rate (pxl/s) (if scanner used)	
Frame rate (if FPA) (frames/s)	12.5 Hz (max. ang. res.), 25 Hz, 50 Hz
Array size (if FPA) (pixels)	N/A
Angular uncertainty (degrees)	Not provided
Angular resolution (horizontal) (degrees)	From 0.125° to 0.5° – depends on frame rate
Color imagery availability/registration (Y/N)	no
Rated operating conditions (such as temperature, ambient brightness, etc.)	-40 °C to +85 °C
Limiting operating conditions	
Data and communication interfaces	Ethernet, CAN
Power supply voltage (V) and consumption (W)	(9 to 27) VDC, 10 W max, 6 W average
Overall size of unit (w x d x h) (mm)	(128x93x85) mm, ECU extra size, weight and power
Weight of complete sensor (kg)	0.9 kg, ECU extra size, weight and power
Retail cost (U.S. $)	Approx. $20K
Lead time to delivery (weeks)	4
What processing software is provided (besides control software)	Visualization software ILV

Manufacturer's Name: Leica		
Product Name and Model Number: C10		
System Architecture: Pulse TOF 3D Laser scanner. All-in-One platform		
Applications: As-built and topographic surveying		
Year Introduced: 2009		
Wavelength of illumination source	(nm)	532
Illumination power	(mW)	1 mW average
Laser safety classification		3R
Beam diameter at exit	(mm)	6 mm
Beam divergence	(mrad)	0.2 mrad
Min./Max. range	(m)	0.1 m to 300 m with 90 % reflective target 0.1 m to 134 m with 18 % reflective target
Range uncertainty (at specified range, reflectivity, and number of measurements)	(mm)	± 4 mm at 50 m, single measurement
Range resolution (depth) [1]	(mm)	< 1 mm at any range
Can system provide range profile (multiple returns) for the same range column	(Y/N)	N
Sensor field of view, horizontal	(degrees)	360°
Sensor field of view, vertical	(degrees)	270°
Pixel or date acquisition rate (if scanner used)	(pxl/s)	50 000
Frame rate (if FPA)	(frames/s)	Variable scan duration
Array size (if FPA)	(pixels)	N/A
Angular uncertainty	(degrees)	0.0034°
Angular resolution	(degrees)	0.0002°
Color imagery availability/registration	(Y/N)	Y, 5 megapixel internal camera
Rated operating conditions (such as temperature, ambient brightness, etc.)		0 °C to +40 °C, full ambient light
Limiting operating conditions		
Data and communication interfaces		Ethernet, Dynamic Internet Protocol (IP) Address, USB 2.0 devices
Power supply voltage (V) and consumption (W)		15 V DC (90 V ac to 260 V ac), 50 W average
Overall size of unit (w x d x h)	(mm)	238 x 358 x 395
Weight of complete sensor	(kg)	13 kg without batteries
Retail cost	(U.S. $)	
Lead time to delivery	(weeks)	
What processing software is provided (besides control software)		

Manufacturer's Name: Leica	
Product Name and Model Number: HDS 6200	
System Architecture: Phase based AM Modulation TOF scanning, 79 m abiguity interval	
Applications: As-built surveys and site surveys, architectural and heritage surveys	
Year Introduced: 2010	
Wavelength of illumination source (nm)	650 to 690
Illumination power (mW)	4.75 mW average
Laser safety classification	3R
Beam diameter at exit (mm)	3 mm at exit
Beam divergence (mrad)	0.22 mrad
Min./Max. range (m)	0.4 m to 79 m
Range uncertainty (mm) (at specified range, reflectivity, and number of measurements)	± 3 mm up to 50 m with 90 % reflectance target ± 5 mm up to 50 m with 18 % reflectance target
Range resolution (depth) [1] (mm)	0.1
Can system provide range profile (multiple returns) for the same range column (Y/N)	N
Sensor field of view, horizontal (degrees)	360°
Sensor field of view, vertical (degrees)	310°
Pixel or date acquisition rate (pxl/s) (if scanner used)	1 016 000 pxl/s
Frame rate (if FPA) (frames/s)	Scan duration at super high resolution – 7 minutes Scan duration at ultra resolution – 27 minutes
Array size (if FPA) (pixels)	N/A
Angular uncertainty (degrees)	0.0071°
Angular resolution (degrees)	0.0018°
Color imagery availability/registration (Y/N)	Y, as third party product
Rated operating conditions (such as temperature, ambient brightness, etc.)	-10 °C to +45 °C, in full light
Limiting operating conditions	
Data and communication interfaces	Ethernet or integrated Wireless LAN (WLAN)
Power supply voltage (V) and consumption (W)	24 V DC (90 V ac to 260V ac), 65 W max.
Overall size of unit (w x d x h) (mm)	199 x 294 x 360
Weight of complete sensor (kg)	14 kg with internal batteries
Retail cost (U.S. $)	
Lead time to delivery (weeks)	
What processing software is provided (besides control software)	

Manufacturer's Name: LightTime		
Product Name and Model Number: MEMScan		
System Architecture: MEMS mirror-based scanning beam, 3D (TOF-range, return beam xy position; plus intensity) image data Capture subsystem (front-end to LADAR imaging system)		
Applications: all forms of transportation vehicle safety, geographic measurements, and military such as Unmanned vehicles imaging, target detection / recognition / acquisition / and surveillance		
Year Introduced: to be introduced in 2011		
Wavelength of illumination source	(nm)	TBD: (860 to 1550) nm
Illumination power	(mW)	TBD
Laser safety classification		Eye Safe / Optional
Beam diameter at exit	(mm)	TBD
Beam divergence	(mrad)	1 mrad
Min./Max. range	(m)	1/100 m at 100 % reflectivity, 1/30 m at 10 % reflect.
Range uncertainty (at specified range, reflectivity, and number of measurements)	(mm)	TBD
Range resolution (depth)	(mm)	TBD
Can system provide range profile (multiple returns) for the same range column	(Y/N)	Yes
Sensor field of view, horizontal	(degrees)	TBD: 32° to 64°
Sensor field of view, vertical	(degrees)	TBD: 24° to 48°
Pixel or date acquisition rate (if scanner used)	(pxl/s)	1M pps (320 x 240 image at 15 frames/s)
Frame rate (if FPA)	(frames/s)	N/A
Array size (if FPA)	(pixels)	N/A
Angular uncertainty	(degrees)	0.1°
Angular resolution	(degrees)	0.1°
Color imagery availability/registration	(Y/N)	Only potentially false color via post-processing
Rated operating conditions (such as temperature, ambient brightness, etc.)		(10 to 40) °C (can be extended), (20 to 80) % RH
Limiting operating conditions		
Data and communication interfaces		TBD
Power supply voltage (V) and consumption (W)		TBD
Overall size of unit (w x d x h)	(mm)	Estimated (100x40x100) mm
Weight of complete sensor	(kg)	<0.5 kg
Retail cost	(U.S. $)	TBD
Lead time to delivery	(weeks)	TBD
What processing software is provided (besides control software)		TBD: Likely a LabView executable

Manufacturer's Name: Mesa Imaging	
Product Name and Model Number: SR4000	

System Architecture: 3D Time-of-flight solid state sensor, 176c144 lock-in pixel array

Applications: Logistics, surveillance, biometry, machine vision, mobile robotics, navigation, mapping

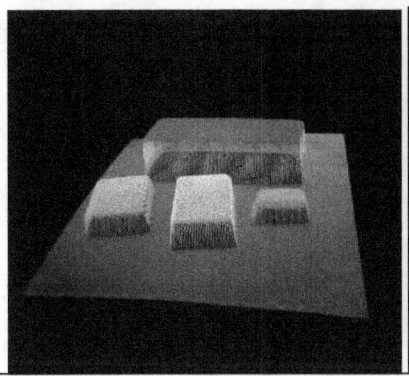 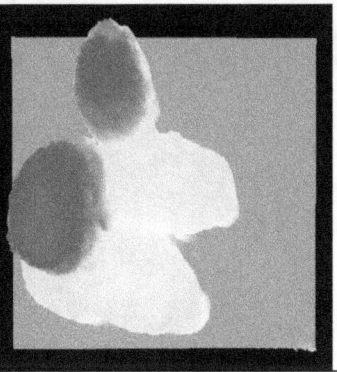 

Year Introduced: 2007

Wavelength of illumination source	(nm)	850
Illumination power	(mW)	0.4
Laser safety classification		1
Beam diameter at exit	(mm)	N/A
Beam divergence	(deg)	+/- 25
Min./Max. range	(m)	0.5 to 10
Range bias (at specified ranges)	(mm)	+/- 1cm over full range
Range resolution (depth)	(mm)	16 bits on max range
Range noise (specify reflectivity of targets and ranges)	(mm rms)	4 mm at 2 m at 30 fps on white target
Sensor field of view, horizontal	(degrees)	43
Sensor field of view, vertical	(degrees)	34
Pixel or date acquisition rate (if scanner used)	(pxl/s)	≈ 500 000
Frame rate (if FPA)	(frames/s)	Up to 54
Array size (if FPA)	(pixels)	176 x 144
Angular resolution	(degrees)	0.25
Angular accuracy	(degrees)	0.25

Ambient temperature (Centigrade) calibrated operating range	0 to 50
Data and communication interfaces	USB2.0 / Ethernet
Power supply voltage (V) and consumption (W)	12 V / 6W
Overall size of unit (w x d x h) (mm)	65 x 65 x 68 (USB) 65 x 65 x 76 (Ethernet)
Weight of complete sensor (kg)	0.47 (USB) 0.51 (Ethernet)
Retail cost (U.S. $)	9000
Lead time to delivery (weeks)	4
Color imagery availability/registration yes/no	No

Manufacturer's Name: Neptec Design Group	
Product Name and Model Number: Laser Metrology System (LMS)	
System Architecture: 3D triangulation with dual-axis autosynchronous laser scanning	
Applications: Dimensional verification, inspection, reverse engineering	
Year Introduced:	
Wavelength of illumination source (nm)	635
Illumination power (mW)	5 mW
Laser safety classification	3R
Beam diameter at exit (mm)	
Beam divergence (mrad)	
Min./Max. range (m)	0.6 m to 5 m
Range uncertainty (mm) (at specified range, reflectivity, and number of measurements)	50 µm @ 0.6 m
Range resolution (depth) (mm)	20 µm (0.6 m), 1.2 mm (5 m)
Can system provide range profile (multiple returns) for the same range column (Y/N)	N
Sensor field of view, horizontal (degrees)	50°
Sensor field of view, vertical (degrees)	30°
Pixel or date acquisition rate (pxl/s) (if scanner used)	10 000
Frame rate (if FPA) (frames/s)	
Array size (if FPA) (pixels)	
Angular uncertainty (degrees)	
Angular resolution (degrees)	10 µm @0.6 m to 73 um @5 m
Color imagery availability/registration (Y/N)	N
Rated operating conditions (such as temperature, ambient brightness, etc.)	15 °C to 40 °C, 20 % to 70 % non-condensing humidity
Limiting operating conditions	
Data and communication interfaces	3/8" – 16 UNC tripod interface, USB 2.0 port
Power supply voltage (V) and consumption (W)	85 to 240 VAC 50/60 Hz
Overall size of unit (w x d x h) (mm)	(330 x 340 x 216) mm (13 x 13.4 x 9.5) in
Weight of complete sensor (kg)	13.6 kg (30 lbs)
Retail cost (U.S. $)	
Lead time to delivery (weeks)	
What processing software is provided (besides control software)	

Manufacturer's Name: Neptec	
Product Name and Model Number: MEMS Lidar Prototype	
System Architecture: MEMS Lidar scanner	

Applications: Canadian Space R&D

Scan of JUNO lunar rover

Year Introduced: TBD		
Wavelength of illumination source	(nm)	1540
Illumination power	(mW)	1500 mW average
Laser safety classification		
Beam diameter at exit	(mm)	2.2 mm
Beam divergence	(mrad)	0.55
Min./Max. range	(m)	3 m / 50 m
Range uncertainty (at specified range, reflectivity, and number of measurements)	(mm)	20 mm to 25 mm (3 m to 50 m) range noise, raw data
Range resolution (depth)	(mm)	<5 mm
Can system provide range profile (multiple returns) for the same range column	(Y/N)	N
Sensor field of view, horizontal	(degrees)	14.7°
Sensor field of view, vertical	(degrees)	13.4°
Scanner pixel or date acquisition rate	(pxl/s)	100 000
Frame rate (if FPA)	(frames/s)	
Array size (if FPA)	(pixels)	
Angular uncertainty	(degrees)	
Angular resolution	(degrees)	
Color imagery availability/registration	(Y/N)	N

Manufacturer's Name: Neptec		
Product Name and Model Number: Obscurant Penetrating Autosynchronous LIDAR - OPAL (prototype)		
System Architecture: Autosynchronous triangulation scanning sensor using pulsed TOF ranging		
Applications: Obstacle detection during helicopter landing		
Year Introduced: Prototype		
Wavelength of illumination source	(nm)	1540 nm
Illumination power	(mW)	Average optical power: 1300
Laser safety classification		Eye safe
Beam diameter at exit	(mm)	10 mm
Beam divergence	(mrad)	0.3 mrad.
Min./Max. range	(m)	Clear air: 2000 m for a 80 % reflective surface 550 m for a 5mm metal wire With obscurants: 50 m for a 80 % reflective surface in a 1 g/m^3 uniform 10 μm dia. dust cloud
Range uncertainty (at specified range, reflectivity, and number of measurements)	(mm)	Production goal: 5 cm at 2 km, 80 % diffusive target, 1 measurement
Range resolution (depth) [1]	(mm)	5 cm
Can system provide range profile (multiple returns) for the same range column	(Y/N)	N
Sensor field of view, horizontal	(degrees)	30°
Sensor field of view, vertical	(degrees)	60°
Pixel or date acquisition rate (if scanner used)	(pxl/s)	Up to 32 000 pixels per second
Frame rate (if FPA)	(frames/s)	N/A
Array size (if FPA)	(pixels)	N/A
Angular uncertainty	(degrees)	Has not been analyzed or measured.
Angular resolution	(degrees)	0.001^o
Color imagery availability/registration	(Y/N)	N
Rated operating conditions (such as temperature, ambient brightness, etc.)		Production goal: -50 °C to +55 °C Prototype: -10 °C to +50 °C
Limiting operating conditions		
Data and communication interfaces		Ethernet, RS-422
Power supply voltage (V) and consumption (W)		28 V, 100 W
Overall size of unit (w x d x h)	(mm)	Production goal: (254 x 254 x 152) Prototype: (457 x 330 x 330)
Weight of complete sensor	(kg)	Production goal: 12 kg Prototype: 19.5 kg
Retail cost	(U.S. $)	
Lead time to delivery	(weeks)	
What processing software is provided (besides control software)		

Manufacturer's Name: Nikon Metrology Inc.	
Product Name and Model Number: Laser Radar Model MV 224 and MV 260	
System Architecture: Frequency Modulated (Chirped 100GHz modulation) Coherent Laser Radar Two Axis Scanner	
Applications: Large volume measurement and inspection of parts , including part-to-CAD comparison, at distances out to 60 m with metrology measurement uncertainty.	
Year Introduced:	
Wavelength of illumination source (nm)	1550
Illumination power (mW)	2 mW
Laser safety classification	US, International Class 1 eye-safe
Beam diameter at exit (mm)	Focused beam 0.16 mm @ 1 m
Beam divergence (mrad)	Diffraction limited approx. 0.06 mrad
Min./Max. range (m)	1 m / 60 m
Range uncertainty (mm) (at specified range, reflectivity, and number of measurements)	0.016 mm at 1 m 0.102 mm at 10 m 0.240 mm at 24 m
Range resolution (depth) [1] (mm)	0.0001 mm
Can system provide range profile (multiple returns) for the same range column (Y/N)	Y
Sensor field of view, horizontal (degrees)	360°
Sensor field of view, vertical (degrees)	+/- 45°
Pixel or date acquisition rate (pxl/s) (if scanner used)	Up to 4000 pixels/s
Frame rate (if FPA) (frames/s)	N/A
Array size (if FPA) (pixels)	N/A
Angular uncertainty (degrees)	
Angular resolution (degrees)	0.000002°
Color imagery availability/registration (Y/N)	Integrated video camera to follow laser line
Rated operating conditions (such as temperature, ambient brightness, etc.)	5 °C to 40 °C, (10 to 90) % humidity (non-condensing)
Data and communication interfaces	
Power supply voltage (V) and consumption (W)	(165 to 270) VAC, 50/60Hz, 5 A max or (85 to 130) VAC, 10 A
Overall size of unit (w x d x h) (mm)	Sensor: (454x381) mm, Workstation: (913x610) mm
Weight of complete sensor (kg)	Sensor: 40 kg, Workstation: 119 kg
Retail cost (U.S. $)	MV 224: $280K, MV 260: $410K
Lead time to delivery (weeks)	8 – 12 weeks
What processing software is provided (besides control software)	

Manufacturer's Name: None (NIST Research)	
Product Name and Model Number: N/A	
System Architecture: Dual frequency combs in a coherent LIDAR setup	
Applications: Very precise absolute distance measurements between two surfaces. Potential (undemonstrated) for 3D imaging if used in frequency multiplexed configuration.	
Year Introduced: 2009	
Wavelength of illumination source (nm)	1550
Illumination power (mW)	9.6 mW or above
Laser safety classification	1 or above
Beam diameter at exit (mm)	approx. 2 mm
Beam divergence (mrad)	approx. 1 mrad
Min./Max. range (m)	dependent on system architecture and target
Range uncertainty (mm) (at specified range, reflectivity, and number of measurements)	approx.±3 µm at 1 µW detected power
Range resolution (depth) [1] (mm)	approx. 2 µm or 3 nm depending on operation
Can system provide range profile (multiple returns) for the same range column (Y/N)	Y
Sensor field of view, horizontal (degrees)	N/A
Sensor field of view, vertical (degrees)	NA
Pixel or date acquisition rate (pxl/s) (if scanner used)	5000 with a single detector – higher if multiplexed
Frame rate (if FPA) (frames/s)	
Array size (if FPA) (pixels)	
Angular uncertainty (degrees)	
Angular resolution (degrees)	
Color imagery availability/registration (Y/N)	
Rated operating conditions (such as temperature, ambient brightness, etc.)	
Limiting operating conditions	
Data and communication interfaces	
Power supply voltage (V) and consumption (W)	
Overall size of unit (w x d x h) (mm)	
Weight of complete sensor (kg)	
Retail cost (U.S. $)	
Lead time to delivery (weeks)	
What processing software is provided (besides control software)	

Manufacturer's Name: PMDTechnologies GmbH	
Product Name and Model Number: PMD[vision] CamCube 2.0	
System Architecture: Modular Demo System with external illumination trigger and exchangeable lenses (CS-mount)	
Applications: Research, Robotics, Under water 3D measurement, Medical Technology and Life Sciences, Safety and Surveillance, Media and Retail, Factory Automation, Consumer and Gaming, Military.	
Year Introduced: 2009	
Wavelength of illumination source (nm)	870 nm (Standard Implementation – other wavelength available upon request) e.g. 525 nm for underwater application
Illumination power (mW)	4000 mW (Standard Implementation – other illumination power available)
Laser safety classification	Class 1 (IEC 60825)
Beam diameter at exit (mm)	
Field of View (°)	40° x 40° (Standard Implementation)
Min./Max. range (m)	0.3 m to 7.5 m (30m with other illumination sources)
Range bias error (mm) (at specified ranges)	
Range resolution (depth) (mm)	
Range noise (specify (mm rms) reflectivity of targets and ranges)	< 3 mm @ 2 m distance, 80 %reflectivity
Sensor field of view, horizontal (degrees)	40 (exchangeable)
Sensor field of view, vertical (degrees)	40 (exchangeable)
Pixel or date acquisition rate (pxl/s) (if scanner used)	parallel acquisition of 200x200 depth data – no scanning
Frame rate (if FPA) (frames/s)	Up to 25 fps, in other applications/demo setups up to 100 fps
Array size (if FPA) (pixels)	204x204
Angular resolution (degrees)	
Angular accuracy (degrees)	
Ambient temperature (Centigrade) calibrated operating range	0 °C to 50 °C
Data and communication interfaces	USB 2.0, Ethernet on request
Power supply voltage (V) and consumption (W)	12V and << 72W
Overall size of unit (w x d x h) (mm)	60 mm x60 mm x60 mm for the camera unit
Weight of complete sensor (kg)	1.4
Retail cost (U.S. $)	11 360 US$
Lead time to delivery (weeks)	Approx. 4 weeks
Color imagery availability/registration (yes/no)	Yes

Manufacturer's Name: PMDTechnologies GmbH	
Product Name and Model Number: PMD[vision] CamCube 3.0	
System Architecture: Modular Demo System with external illumination trigger and exchangeable lenses (CS-mount)	
Applications: Research, Robotics, Under water 3D measurement, Medical Technology and Life Sciences, Safety and Surveillance, Media and Retail, Factory Automation, Consumer and Gaming, Military.	
Year Introduced: 2010	
Wavelength of illumination source (nm)	870 nm (Standard Implementation – other wavelength available upon request) e.g. 525 nm for underwater application
Illumination power (mW)	4000 mW (Standard Implementation – other illumination power available)
Laser safety classification	Class 1 (IEC 60825)
Beam diameter at exit (mm)	5 mm
Field of View (°)	40° x 40° (Standard Implementation)
Min./Max. range (m)	0.3 m to 7.5 m (30 m with other illumination sources)
Range bias error (mm) (at specified ranges)	8 mm @ 4 m distance, 75 %reflectivity
Range resolution (depth) (mm)	Equal to range noise
Range noise (specify (mm rms) reflectivity of targets and ranges)	< 3 mm @ 4 m distance, 75 %reflectivity
Sensor field of view, horizontal (degrees)	40 (exchangeable)
Sensor field of view, vertical (degrees)	40 (exchangeable)
Pixel or date acquisition rate (pxl/s) (if scanner used)	parallel acquisition of 200x200 depth data – no scanning
Frame rate (if FPA) (frames/s)	Up to 40 fps, in other applications/demo setups up to 100 fps with Region-of-Interest function
Array size (if FPA) (pixels)	200x200
Angular resolution (degrees)	0.2°
Angular accuracy (degrees)	± 0.01°
Ambient temperature (Centigrade) calibrated operating range	0 °C to 50 °C
Data and communication interfaces	USB 2.0, Ethernet on request
Power supply voltage (V) and consumption (W)	12 V and << 72W
Overall size of unit (w x d x h) (mm)	60 mm x60 mm x60 mm for the camera unit
Weight of complete sensor (kg)	1.4
Retail cost (U.S. $)	11,360 US$
Lead time to delivery (weeks)	Approx. 4 weeks
Color imagery availability/registration (yes/no)	Yes

Manufacturer's Name: Riegl GmbH	
Product Name and Model Number: VZ 1000	
System Architecture: Pulse Time of Flight Scanning LIDAR	
Applications: Civil Engineering, BIM/Architecture, Archeology, City Modeling, Topography & Mining, As Builts, Mobile data acquisition	
Year Introduced: 2010	
Wavelength of illumination source (nm)	1550
Illumination power (mW)	
Laser safety classification	Class 1 eye-safe
Beam diameter at exit (mm)	7 mm at exit, 30 mm at 100 m
Beam divergence (mrad)	0.3 mrad
Min./Max. range (m)	2.5 m/ 1400 m
Range uncertainty (mm) (at specified range, reflectivity, and number of measurements)	±8 mm at max. range
Range resolution (depth) [1] (mm)	3 mm
Can system provide range profile (multiple returns) for the same range column (Y/N)	Yes. Online wave form processing 1 500 000 target returns per second
Sensor field of view, horizontal (degrees)	360°
Sensor field of view, vertical (degrees)	100°
Pixel or date acquisition rate (pxl/s) (if scanner used)	122 000 maximum measurement rate
Frame rate (if FPA) (frames/s)	60° max. per second (horizontal)
Array size (if FPA) (pixels)	N/A
Angular uncertainty (degrees)	0.005°
Angular resolution (degrees)	0.0005°
Color imagery availability/registration (Y/N)	Yes/Yes
Rated operating conditions (such as temperature, ambient brightness, etc.)	0 °C to +40 °C, 100 % humidity, outdoor ambient light
Limiting operating conditions	
Data and communication interfaces	TCP/IP, USB, Wireless
Power supply voltage (V) and consumption (W)	(11 to 32) VDC, 82 W typical
Overall size of unit (w x d x h) (mm)	308 x 200
Weight of complete sensor (kg)	9.8 kg
Retail cost (U.S. $)	
Lead time to delivery (weeks)	
What processing software is provided (besides control software)	

Manufacturer's Name: Riegl GmbH		
Product Name and Model Number: VZ 400		
System Architecture: Pulse Time of Flight Scanning LIDAR		
Applications: Civil Engineering, BIM/Architecture, Archeology, City Modeling, Topography, As Builts, Mobile data acquisition		
Year Introduced: 2008		
Wavelength of illumination source	(nm)	1550
Illumination power	(mW)	1mW
Laser safety classification		Class 1 eye-safe
Beam diameter at exit	(mm)	7 mm at exit, 30 mm at 100m
Beam divergence	(mrad)	0.3 mrad
Min./Max. range	(m)	1.5 m/500 m
Range uncertainty (at specified range, reflectivity, and number of measurements)	(mm)	5 mm at 500 m
Range resolution (depth) [1]	(mm)	3 mm one sigma 100 m range at Riegl Test range
Can system provide range profile (multiple returns) for the same range column	(Y/N)	Yes. Online wave form processing. 1 500 000 target returns per second
Sensor field of view, horizontal	(degrees)	(0 to 360) °
Sensor field of view, vertical	(degrees)	100
Pixel or date acquisition rate (if scanner used)	(pxl/s)	125 000 max. meas./s
Frame rate (if FPA)	(frames/s)	60° max. per second (horizontal)
Array size (if FPA)	(pixels)	N/A
Angular uncertainty	(degrees)	0.005°
Angular resolution	(degrees)	0.0005°
Color imagery availability/registration	(Y/N)	Yes/yes
Rated operating conditions (such as temperature, ambient brightness, etc.)		0 °C to 40 °C, 100% humidity, outdoor ambient light
Limiting operating conditions		
Data and communication interfaces		TCP/IP, USB, Wireless
Power supply voltage (V) and consumption (W)		(11 to 32)VDC, 65 W
Overall size of unit (w x d x h)	(mm)	308 x 180
Weight of complete sensor	(kg)	9.8 kg
Retail cost	(U.S. $)	$125 000
Lead time to delivery	(weeks)	6
What processing software is provided (besides control software)		RiScan Pro

Manufacturer's Name: TetraVue		
Product Name and Model Number: TetraCorder		
System Architecture: Real-time high resolution, FPA-based 3D camera, standard CCD/CMOS arrays		
Applications: construction, infrastructure monitoring, structural analysis, biometrics, forensics, video games, quality assurance		
Year Introduced: expected in 2011 or 2012		
Wavelength of illumination source	(nm)	IR
Illumination power	(mW)	
Laser safety classification		Class 1
Beam diameter at exit	(mm)	
Beam divergence	(mrad)	
Min./Max. range	(m)	1 to 100
Range uncertainty (at specified range, reflectivity, and number of measurements)	(mm)	
Range resolution (depth) [1]	(mm)	1 mm RMS @20 m, single frame
Can system provide range profile (multiple returns) for the same range column	(Y/N)	No
Sensor field of view, horizontal	(degrees)	30
Sensor field of view, vertical	(degrees)	30
Pixel or date acquisition rate (if scanner used)	(pxl/s)	
Frame rate (if FPA)	(frames/s)	30
Array size (if FPA)	(pixels)	> 2 M pixel
Angular uncertainty	(degrees)	< .03
Angular resolution	(degrees)	< .03
Color imagery availability/registration	(Y/N)	Y
Rated operating conditions (such as temperature, ambient brightness, etc.)		Any ambient lighting, including full sun
Limiting operating conditions		
Data and communication interfaces		
Power supply voltage (V) and consumption (W)		
Overall size of unit (w x d x h)	(mm)	300 x 200 x 150
Weight of complete sensor	(kg)	< 8
Retail cost	(U.S. $)	
Lead time to delivery	(weeks)	
What processing software is provided (besides control software)		

Manufacturer's Name: Trimble		
Product Name and Model Number: CX Scanner		
System Architecture: Combined Pulse & Phase TOF Laser Scanner		
Applications: Civil surveying, building information modeling (BIM), heritage documentation, and forensic and accident investigation		
Year Introduced: 2010		
Wavelength of illumination source	(nm)	660
Illumination power	(mW)	
Laser safety classification		3R
Beam diameter at exit	(mm)	3 mm at exit, 8 mm at 25 m, 13 mm at 50 m
Beam divergence	(mrad)	0.2 mrad
Min./Max. range	(m)	1 m to 80 m
Range uncertainty (at specified range, reflectivity, and number of measurements)	(mm)	± 2 mm
Range resolution (depth) [1]	(mm)	< 1 mm
Can system provide range profile (multiple returns) for the same range column	(Y/N)	N
Sensor field of view, horizontal	(degrees)	360°
Sensor field of view, vertical	(degrees)	300°
Pixel or date acquisition rate (if scanner used)	(pxl/s)	54 000
Frame rate (if FPA)	(frames/s)	Max. horizontal points per 360°, 180 000
Array size (if FPA)	(pixels)	N/A
Angular uncertainty	(degrees)	0.004^o H and 0.007^o V
Angular resolution	(degrees)	0.002^o
Color imagery availability/registration	(Y/N)	Y, video
Rated operating conditions (such as temperature, ambient brightness, etc.)		$0\ ^oC$ to $+40\ ^oC$
Limiting operating conditions		
Data and communication interfaces		USB Flash drive, data transfer cable, WLAN antenna
Power supply voltage (V) and consumption (W)		24 V DC (90 to 240 V ac), 50 W typical
Overall size of unit (w x d x h)	(mm)	120 x 520 x 355
Weight of complete sensor	(kg)	11.8 kg
Retail cost	(U.S. $)	
Lead time to delivery	(weeks)	
What processing software is provided (besides control software)		

Manufacturer's Name: Velodyne Acoustics, Inc.	
Product Name and Model Number: High Definition LIDAR – HDL 32E	
System Architecture: Multiple laser/multiple plane pulse TOF real-time LIDAR Pulse duration: 5 ns, APD detectors, entire sensor rotates 360°	
Applications: Obstacle detection and avoidance for autonomous vehicle navigation systems Mobile surveying/mapping Industrial uses	
Year Introduced: 2010	
Wavelength of illumination source (nm)	905 nm
Illumination power (mW)	
Laser safety classification	Class I – eye safe
Beam diameter at exit (mm)	
Beam divergence (mrad)	
Min./Max. range (m)	5 cm to 100 m
Range uncertainty (mm) (at specified range, reflectivity, and number of measurements)	< ± 20 mm
Range resolution (depth) [1] (mm)	< 20 mm
Can system provide range profile (multiple returns) for the same range column (Y/N)	Yes
Sensor field of view, horizontal (degrees)	360°
Sensor field of view, vertical (degrees)	40°
Pixel or date acquisition rate (pxl/s) (if scanner used)	Up to 800k pxl/s
Frame rate (if FPA) (frames/s)	(5 to 20) Hz FOV update
Array size (if FPA) (pixels)	N/A
Angular uncertainty (degrees)	Not provided
Angular resolution (degrees)	1.25° in vertical
Color imagery availability/registration (Y/N)	No, but provides intensity image
Rated operating conditions (such as temperature, ambient brightness, etc.)	-40 °C to +85 °C
Limiting conditions	
Data and communication interfaces	100 MBPS UDP Ethernet packets
Power supply voltage (V) and consumption (W)	12 V (9 VDC to 32 VDC) @ 2 A
Overall size of unit (w x d x h) (mm)	150 mm tall cylinder x 86 mm OD dia.
Weight of complete sensor (kg)	< 2 kg
Retail cost (U.S. $)	Not provided
Lead time to delivery (weeks)	Accepting orders
What processing software is provided (besides control software)	

Manufacturer's Name: Velodyne Acoustics, Inc.	
Product Name and Model Number: High Definition LIDAR – HDL 64E S2	
System Architecture: Multiple laser/multiple plane pulse TOF real-time LIDAR Pulse duration: 5 ns, APD detectors, entire sensor rotates 360°	
Applications: Obstacle detection and avoidance for autonomous vehicle navigation systems Mobile surveying/mapping Industrial uses	
Year Introduced: 2008	
Wavelength of illumination source (nm)	905 nm
Illumination power (mW)	Max peak power: 60 W, automatic power control
Laser safety classification	Class I – eye safe
Beam diameter at exit (mm)	6.4 mm at 10 m
Beam divergence (mrad)	0.13 mrad
Min./Max. range (m)	50 m for pavement – 120 m for cars & foliage
Range uncertainty (mm) (at specified range, reflectivity, and number of measurements)	< ±20 mm
Range resolution (depth) [1] (mm)	<20 mm
Can system provide range profile (multiple returns) for the same range column (Y/N)	N
Sensor field of view, horizontal (degrees)	360°
Sensor field of view, vertical (degrees)	26.8°
Pixel or date acquisition rate (pxl/s) (if scanner used)	>1.333 M pxl/s
Frame rate (if FPA) (frames/s)	(5 to 15) Hz FOV update
Array size (if FPA) (pixels)	N/A
Angular uncertainty (degrees)	Not provided
Angular resolution (degrees)	0.09° in azimuth, 0.41° in vertical
Color imagery availability/registration (Y/N)	No, but provides intensity image
Rated operating conditions (such as temperature, ambient brightness, etc.)	-10 °C to +50 °C
Limiting conditions	
Data and communication interfaces	100 MBPS UDP Ethernet packets
Power supply voltage (V) and consumption (W)	12 V (16 V Max) @ 4 A
Overall size of unit (w x d x h) (mm)	254 mm tall cylinder x 203 mm OD dia.
Weight of complete sensor (kg)	< 15 kg
Retail cost (U.S. $)	$75K
Lead time to delivery (weeks)	2 – 4 weeks
What processing software is provided (besides control software)	

Manufacturer's Name: Zoller + Froehlich GmbH		
Product Name and Model Number: Imager 5010		
System Architecture: Stand alone phase-based laser scanner		
Applications: very high speed and long-range (187 m) laser scanning Industrial plants, forestry, forensics, security & accident documentation		
Year Introduced: 2010		
Wavelength of illumination source	(nm)	1500
Illumination power	(mW)	
Laser safety classification		Class 1
Beam diameter at exit	(mm)	≈3.5 mm (at 0.1 m distance)
Beam divergence	(mrad)	< 0.3 mrad
Min./Max. range	(m)	(0.3 to 187.3) mm (ambiguity interval)
Range uncertainty (at specified range, reflectivity, and number of measurements)	(mm)	< ±1 mm
Range resolution (depth) [1]	(mm)	0.1 mm
Can system provide range profile (multiple returns) for the same range column	(Y/N)	N
Sensor field of view, horizontal	(degrees)	360°
Sensor field of view, vertical	(degrees)	320°
Pixel or date acquisition rate (if scanner used)	(pxl/s)	Up to 1,016,027 pxl/s 127 000 pxl/s for typical resolution scan
Frame rate (if FPA)	(frames/s)	26 s/frame to > 1h/frame (for extremely high res)
Array size (if FPA)	(pixels)	1250 x 1250 preview res., 40k x40k ultrahigh res. 10k x 10k high res., 100k x 100k extreme res.
Angular uncertainty	(degrees)	0.007° rms
Angular resolution	(degrees)	< 0.0004°
Color imagery availability/registration	(Y/N)	Y
Rated operating conditions (such as temperature, ambient brightness, etc.)		-10 °C to +45 °C
Limiting operating conditions		
Data and communication interfaces		Ethernet, WLAN, 2xUSB2.0, LEMO 9/7
Power supply voltage (V) and consumption (W)		24 V dc (100 V ac to 240 V ac) < 65W
Overall size of unit (w x d x h)	(mm)	170 x 286 x 395
Weight of complete sensor	(kg)	9.8 kg
Retail cost	(U.S. $)	
Lead time to delivery	(weeks)	
What processing software is provided (besides control software)		

Manufacturer's Name: Zoller + Froehlich GmbH		
Product Name and Model Number: Imager5006-I (upgrade version of Imager5006)		
System Architecture: Stand alone phase-based laser scanner, user interface, PC+HDD, battery integrated. Triple-frequency AMCW		
Applications: very high speed, mid-range (0.4 m to 79 m) laserscanning		
Year Introduced: end of 2008 (Imager5006: end of 2006)		
Wavelength of illumination source	(nm)	visible (635 to 690) nm
Illumination power	(mW)	up to 29 mW
Laser safety classification		3R
Beam diameter at exit	(mm)	3 mm
Beam divergence	(mrad)	0.22 mrad
Min./Max. range	(m)	0.4 m to 79 m (ambiguity interval)
Range accuracy (at specified ranges)	(mm)	< 1 mm, ranges up to 50m
Range resolution (depth)	(mm)	0.1 mm
Range noise (with 20 % or other reflectivity target at specified ranges) (all specs are mm rms, at 127.000 pixel/s data rate, high-power mode)	(mm rms)	at 10m: 100 %: 0.4mm / 20 %: 0.7mm / 10 %: 1.2mm at 25m: 100 %: 0.7mm / 20 %: 1.5mm / 10 %: 2.6mm at 50m: 100 %: 1.8mm / 20 %: 3.5mm / 10 %: 6.6mm
Sensor field of view, horizontal	(degrees)	360°
Sensor field of view, vertical	(degrees)	310°
Pixel or date acquisition rate (if scanner used)	(pxl/s)	up to 500 000 pxl/s, 127 000 / 254 000 typical
Frame rate (if FPA)	(frames/s)	25 s/frame … 26:40 min/frame
Array size (if FPA)	(pixels)	depending on selected resolution: 1250 x 1250 ("preview" resolution) up to 40.000 x 40.000 ("ultrahigh" resolution), 10.000 x 10.000 ("high" typical)
Angular resolution	(degrees)	0.0018° hor/vert
Ambient temperature calibrated operating range	(Centigrade)	-10 °C … +45 °C operation 0.4 m … 78 m (ambiguity)
Data and communication interfaces		Ethernet, WLAN, 2xUSB2.0, digital RS232 and I/O for GPS, IMU connection, connectors for motorized digital color camera
Power supply voltage (V) and consumption (W)		24 V dc (ext. power supply: (90 V ac to 260 V ac) / max. 65W
Overall size of unit (w x d x h)	(mm)	(286 x 190 x 372) mm
Weight of complete sensor	(kg)	14 Kg (including battery)
Retail cost	(U.S. $)	approx. 109 K $
Lead time to delivery	(weeks)	2 – 3 weeks

Appendix C: 3D Imaging Software Packages

Table C. 1. List of Software Packages for 3D Imaging Systems.

Manufacturer	Product Name
3rdTech	SceneVision-3D
Certainty3D, LLC	TopoDOT
InnovMetric Software Inc	PolyWorks V11
kubit USA	PointCloud 5
Leica Geosystems	Leica Cyclone Family of Software [1]
Leica Geosystems	Leica CloudWorx for AutoCAD (Basic and Pro versions)
Leica Geosystems	Leica CloudWorx for MicroStation
Leica Geosystems	Leica CloudWorx for PDMS
Leica Geosystems	Leica CloudWorx for Intergraph SmartPlant Review
Leica Geosystems	Leica TruView FREE Web Viewer
Leica Geosystems	Leica Cyclone II TOPO
Leica Geosystems Inc.	Cyclone
Maptek I-Site 3D Laser Imaging	I-Site Studio 3.3
Maptek I-Site 3D Laser Imaging	I-Site Forensic 2.1
Maptek I-Site 3D Laser Imaging	I-Site Voidworks 2.0
Riegl	RiSCAN PRO
Riegl USA	Phidias
Riegl USA	RiScan PRO
Spatial Integrated Systems Inc	3 DIS - 3 Dimensional Imaging & Scanning
Topcon Corporation	ScanMaster
Topcon Positioning Systems	ScanMaster
Trimble	Trimble RealWorks
Trimble	LASERGen
Z+F UK LTD	LFM Software
Geomagic	Geomagic Studio
3DReshaper	3DReshaper
New River Kinematics	Spatial Analyzer
Delcam	CopyCAD
Terrasolid	TerraScan
Pointools Ltd	Pointools
Rapidform	XOR/Redesign, XOV/Verifier, XOS/Scan
VirtualGeo	CloudCube
Gexel	JRC 3D Reconstructor
Virtual Grid	VRMesh
QuantaPoint	QuantaCAD
NOTE: Shaded cells are software packages in POB survey.	

The responses from the software companies not in the POB survey are given in Table C. 2. The two questions in the shaded cells were questions added to the POB questions.

Table C. 2. Responses for some of the software packages not in the POB survey.

Manufacturer	3DReshaper	Gexcel	New River Kinematic	Virtual Grid	Pointools	QuantaPoint
Product Name	3DReshaper	JRC 3D Reconstructor	Spatial Analyzer	VRMesh	Pointools Edit	QuantaCAD
Operating systems supported (if one is preferred, please state, is fully 64bit supported)	32 bits but 64 bits in 2011	Windows (98, XP, Vista 7), it works with 32- or 64 bit system	Windows XP, Vista, & 7 (32-bit and 64-bit). Windows 7 preferred. Runs fine as a 32-bit application on 64-bit operating systems.	Microsoft Windows 7, Vista and XP (32 and 64 bit) are fully supported.	Windows XP, Vista and 7, both 32bit and 64bit builds are in release	Windows XP (32-bit), Windows 7 (32-bit and 64-bit)
Minimum CPU requirement	less than a 2 year old Dual core	Pentium 4 or equivalent, with SSE2 extensions	No minimum requirement, but 3GHz or faster Dual-Core processor is recommended.	Intel / AMD CPU 1.8 GHz	Pentium 4 3.0 GHz or better	800 MHz Pentium III Class
Minimum RAM required	2 GB	512MB RAM, 1 GB recommended	No minimum requirement (512 MB is probably the practical minimum), but 8 GB of RAM (or more if a 64-bit OS) is recommended.	512 MB RAM	1 GB + recommended	512 MB
Space required on hard disk to properly run application, including swap space, etc. (list in Mb)	2 GB	512 MB to run the software, typical project dimension 2GB or more	500 MB	10.0 GB free disk space	200 MB	20 MB

147

Manufacturer	3DReshaper	Gexcel	New River Kinematic	Virtual Grid	Pointools	QuantaPoint
Other hardware requirements	No	graphic card: compatible OpenGL 1.5 (minimum). Suggested configuration OpenGL 2+, (i.e. NVIDIA- GForce 6 or major).	1024x768 video driver.	3-button, scroll-wheel mouse. A 3D accelerated graphics card that supports OpenGL	OpenGL graphics card, min 256Mb with OpenGL 1.4+ support	Support for DirectX 8.1, Microsoft .NET Framework 2.0, Latest Drivers for the Graphics Adapter
Cloud Editing/Analysis						
Can features be defined with user-created code libraries?	No yet, otherwise in C++	no (the user can run some batch processing sequences)	Yes	No	No	No
Feature codes exportable to CAD software? (specify which software)	No	DXF export supported; Direct link with AutoCAD through PointCloud (by Kubit)	No	No	No	No
Can user compare cloud or shapes fitted to clouds to plan or perform theoretical shape and interference checking? (State which, all or none.)	Yes all	A tool called Inspection allows to compare mesh model (from theoretical or measured models) and point clouds. A numerical and graphical map of the model differences is produced	Yes-Both	Yes for all	No	Yes

148

Manufacturer	3DReshaper	Gexcel	New River Kinematic	Virtual Grid	Pointools	QuantaPoint
Ability to make measurements such as distances, angles, areas, volumes, of lines, planes, shapes and other surfaces from cloud? (State which, all or none.)	Yes all	linear, angular, volume, area are supported	Yes-All	Yes for all	distance, angles	Yes
Can user overlay or drape a photograph from an external source (e.g., digital camera) on cloud or elements extracted from cloud?	Yes but only on a meshed model, not onto cloud.	Yes. The user can calibrate any external camera; the calibration is based on range scans or generic 3D points. The texture mapping tool allows: i)real time photo projection on the 3D model, ii) mapping of the texture onto the triangulated 3D model; iii) mosaic and blending of calibrated photos; iv) texture onto triangulated 3D model or point cloud using camera mounted on laser scanner.	No	No	No	Yes
Ability to register scans without the use of targets?	Yes	Yes	Yes	Yes, VRMesh has the ability to register scans without/with the use of targets.	No	Yes (if converted from registered laser data).

Manufacturer	3DReshaper	Gexcel	New River Kinematic	Virtual Grid	Pointools	QuantaPoint
Ability to place several clouds from different scans in coordinated 3D space using total station or GPS survey data that has been used to determine positions of scanner and alignment of scans?	Just with XYZ coordinates, manually entered.	Yes	Yes	No. But VRMesh can place several clouds from different scans in coordinated 3D space using the marked points.	Yes	Yes
Analyze points in a cloud representing shapes such as planes, cylinders and spheres to detect measurement outliers?	Yes	Yes (Planes, Sphere and Cylinder can be created by point cloud)	Yes	No	No	Yes
Ability to integrate scans with floor plans, engineering drawings of objects and surveyed information? (State which, all or none.)	Certain entities can be imported like polylines; not all.	Ability to integrate scans (virtual scan tool) from predefined user view (orthographic, perspective and cylindrical). Mesh model can be imported and integrated with point clouds.	Yes-surveyed information and engineering drawings.	No	Yes, integrate with 3d models or 3d/2d linework	Yes
Automate decimation of points in selectable areas to make data files as compact as possible?	Yes but on the entire cloud bases on a constant density	Yes	Yes	Yes	Yes	No

Manufacturer	3DReshaper	Gexcel	New River Kinematic	Virtual Grid	Pointools	QuantaPoint
Is fitting of lines, planes and shapes to cloud done manually or automatically, or both?	Both	Manually	Manually	Manually	No	Automatically
- For automatic and manual fitting, what techniques are used or available (e.g. least squares, taking average, etc.)?	Least squares method	Least squares	Least squares	Least squares	No	Proprietary
Are standard deviations of fitted parameters provided? (e.g., std of fitted plane parameters - not residual value of error function)**	Standard deviation	Std of fitted plane is provided	Yes	The standard deviations will be shown	no	Yes.
Ability to automatically track lines or limits of areas by color or texture discrimination?	Not really. Points can be automatically explored according to a specific color.	No	No	No	Yes	No
Ability to automatically calculate and list alignment of center line of shapes (such as a pipe) containing straight and curved segments such as elbows?	Yes but not every time according to the model.	No	No	No	No	Yes

Manufacturer	3DReshaper	Gexcel	New River Kinematic	Virtual Grid	Pointools	QuantaPoint
Maximum number of points that can be loaded	Depending on your computer but there is no real limit	50 million on 32 bit, 80 million on 64 bit	Many millions (depends on hardware--RAM mostly)	More than 1 billion	10 billion +	Only limited by computer memory
Automatic removal of noise (e.g., cars on road, vegetation, etc.)?	Yes - aberrant points and according to a distance parameter	Yes (set of point filters, i) median filter, ii) strain point, iii) inclination, etc)	Yes (depends on the case)	Yes. In the coming version 5.x, VRMesh can automatically detect and remove vegetation, building, etc. to generate bare-earth surface from point clouds	Not auto	Yes (user-defined)
Rendering/CAD Model Generation/Viewing						
Does software automatically or manually generate or create CAD models or model segments from point clouds and other known information? (Specify level of automation and intelligence.)	Mostly manually but it will tend to more automatic shape recognition	Manually	Manually --can deform a nominal CAD surface using scanned data	VRMesh 5.0 has the tools and power to clean up and simplify large-sized point clouds, generate high-quality triangulated meshes, and recreate NURBS surfaces for further manipulation in your CAE and CAD applications	No	Augmented remodeling
Are items (CAD models such as pipes, steel, flanges, elbow) fit to the point cloud using standard object tables/catalogs?	No	No	No	NO	No	Yes, nominal specifications are used.

152

Manufacturer	3DReshaper	Gexcel	New River Kinematic	Virtual Grid	Pointools	QuantaPoint
Create statistical quality assurance reports on the modeled objects?	Yes	No	Yes	No	No	No
Automatically compute, without user interaction, a full 3D polygonal mesh (not view-based) from a point cloud?	Yes	No (view dependent	No	Yes	No	No
Perform contour generation?	Yes	Yes, based on cross section	No	Yes	No	No
Perform volume calculation capabilities?	Yes	Yes	No	Yes	No	No
Perform solid modeling (volume generation) based on user-defined lines, planes and other surfaces as bounds?	Yes. Solid modeling is based on the creation of a network of curves with intersection.	Yes, based on manual points selection	No	No	No	For pipes and structure.
Perform profile and cross-section generation along any cutting plane, family of planes or road alignment?	Yes, planar, radial, customizable, along a curve...	Yes, according to planes and polylines	Yes	Yes	No	Yes
Have edge detection technology to determine boundaries of solids, planes and other shapes?	Yes : feature line extraction	Yes, angular and depth edge detection	No	Yes	No	Yes

153

Manufacturer	3DReshaper	Gexcel	New River Kinematic	Virtual Grid	Pointools	QuantaPoint
Perform automatic extraction of standard shapes from cloud (e.g. pipe fittings, structural steel members, etc.)?	No	Only, plane, sphere and cylinder	No	NO	No	Yes
Can user view cloud or generated shapes or models from any viewpoint in 3D?	Yes	Yes	Yes	Yes	Yes	Yes
Are fly-throughs or walk-throughs supported?	Neither one nor the other	Yes	No	Not fully supported	Yes	Yes
Have intelligent display of detail depending on scale of the view?	No	Yes	No	No	Yes	Yes
Can user select transparent/opaque surface for cloud and CAD shapes?	Yes	No	Yes	Yes		For CAD.
Which export formats are supported?	See on http://www.3dreshaper.com/en1/En_faq.htm	Dxf, txt, ptx, ptc, wrml, collada	ASCII, STEP, IGES, VDA, SAT, DXF, POL, STL, AIMS TDF, Imageware Cloud File	Import: vrg, stl, ply, txt, asc, xyz, las, ptx, pts, 3dd, pcp, dxf, obj, 3ds, wrl, byu, vtk, tif, jpg, png Export: vrg, stl, txt, asc, las, pts, igs, dxf, wrl, obj, 3ds, ply, rib, iv, byu, vtk, tif, jpg, png	All major scanning formats (pts, xyz, pts, ptx, ptg, ptz, fls, fws, 3dd, rxp, rsp, cl3, las, bin + others)	PTX, PTS.

Manufacturer	3DReshaper	Gexcel	New River Kinematic	Virtual Grid	Pointtools	QuantaPoint
Specify other measurement tools (e.g., clearance, cut/fill, table of elevation differences)		Scan differences in time; point classification according to depth, inclination, overlapped color layers	Querying points to points, line/plane angles, deviation from CAD, orientation to orientation, distance (incorporating measurement uncertainty), Unified Spatial Metrology Network (incorporating uncertainty characteristics of coupled measurement devices).	Remove redundant points, remove isolated points, clip by plane, cut by lasso, etc.		Point, Distance, Cylinder, I-Beam, Column, Plane, Wall, Ceiling, Door
Can the pointcloud be rendered with visualization effects (e.g., intensity mapping, elevation mapping, shading, silhouette)?	Yes	Yes	No	The point cloud can be rendered with elevation mapping in V5.0. In the next version V5.x, the intensity mapping will be added.	Yes	Yes
Can the software automatically detect scan targets?	No	No	No	In the next version v5.x, VRMesh can automatically detect scan targets, e.g., tree, building.	No	Yes
Miscellaneous						
Provide high-speed thumbnail views of scans, clouds, photographic images and generated shapes?	Texturing : images applied on a meshed model	No	No		No	Yes

155

Manufacturer	3DReshaper	Gexcel	New River Kinematic	Virtual Grid	Pointools	QuantaPoint
Can client/server system support multiple users?	Yes	Yes	No	No	No	Yes
Is client/server system supported to enable several clients contributing to a single project?	No	No	No	No	No	Yes
Internal calculations in single or double precision ?**	Double precision	Double	Double	Double precision	Double	Double

** Question was not part of the POB survey question.

REFERENCES

1. Stone, W.C., et al., *Performance Analysis of Next-Generation LADAR for Manufacturing, Construction, and Mobility*. 2004, NISTIR 7117, .

2. ASTM, *Standard Terminology for Three-Dimensional (3D) Imaging Systems E2544-10* 2010, ASTM International: West Conshohocken, PA.

3. US Army Corps of Engineers, *Terrestrial 3D Laser Scanners*, Dept. of Army, Editor. 2007.

4. Beraldin, J.A. *Basic Theory on Surface Measurement Uncertainty of 3D Imaging Systems*. in *Proceedings of SPIE-IS&T Electronic Imaging*. 2008.

5. Montagu, J. and H. DeWeerd, eds. *Optomechanical Scanning Applications, Techniques, and Devices*. Infrared & Electro-Optical Systems Handbook, ed. W.D. Rogatto. Vol. 3. 1993, SPIE.

6. Schwarte, R., ed. *Principles of 3-D Imaging Techniques*. Handbook of Computer Vision and Applications. 1999, Academic Press.

7. Lange, R., *Time-of-Flight Distance Measurement with Custom Solid-State Image Sensors in CMOS/CCD-Technology*, in *Doctoral Dissertation*. 2000, Dept. of Electrical Engineering and Computer Science at University of Siegen, Germany

8. Breuckmann, B., *Bildverarbeitung und optische Messtechnik in der industriellen Praxis*. 1993, Munchen: Francis-Verlag.

9. Dorsch, R.G., G. Hausler, and J.M. Hermann, *Laser triangulation: fundamental uncertainty in distance measurement*. Applied Optics, 1994. **33**(7): p. 1306 - 1314.

10. Besl, P., *Active Optical Range Imaging Sensors*. Machine Vision and Applications, 1988. **1**: p. 127 - 152.

11. Schwarte, R., *Pseudo-noise (PN) - laser radar without scanner for extremely fast 3D-imaging and navigation*, in *MIOP*. 1997: Stuttgart.

12. Hecht, E. and A. Zajac, *Optics*. 1974: Addison-Wesley.

13. Zimmermann, E., Y. Salvadé, and R. Dändliker, *Stabilized three-wavelength source calibrated by electronic means for high-accuracy absolute distance measurements*. Optics Letters, 1996. **21**(7): p. 531 - 533.

14. Engelhardt, K., *Methoden und Systeme der optischen 3-D Messtechnik*, in *XI. Internationaler Kurs fur Ingenieurvermessung*. 1992, Ferd. Dummlers Verlag: Zurich.

15. Bourquin, S., *Low-coherence interferometry based on customized detector arrays*, in *PhD Dissertation 2171*. 2000, EPFL-Lausanne, Switzerland.

16. Haussecker, J.B. and P. Giessler, in *Handbook of Computer Vision and Applications* 1999, Academic Press: San Diego.

17. Schmitt, J.M., *Optical coherence tomography (OCT): A review*. IEEE J.Quantum Electron., 1999. **5**: p. 1205-1215.

18. Beer, S., *Real-Time Photon-Noise Limited Optical Coherence Tomography Based on Pixel-Level Analog Signal Processing*, in *Doctoral Dissertation*. 2006, Faculty of Science at the University of Neuchatel, Switzerland.

19. Mayer, R., *Scientific Canadian: Invention and Innovation From Canada's National Research Council*. 1999, Vancouver: Raincoats Books.

20. Rioux, M., *Laser range finder based on synchronized scanners*. Applied Optics, 1994. **23**(21): p. 3837-3844.

21. Beraldin, J.A., et al., *Registered intensity and range imaging at 10 mega-samples per second.* Optical Engineering, 1992. **31**(1): p. 88-94.

22. Kramer, J., *Photo-ASICs: Integrated Optical Metrology Systems with Industrial CMOS Technology*, in *PhD Dissertation 10186, .* 1993, ETH Zurich, Switzerland.

23. Beraldin, J.A., et al., *Active 3D Sensing*, in *Modelli E Metodi per lo studio e la conservazione dell'architettura storica.* 2000: Scola Normale Superiore Pisa 10. p. 22-46.

24. Beraldin, J.A., et al. *Optimized Position Sensors for Flying-Spot Active Triangulation Systems.* in *IEEE 4th International Conference on 3D Digital Imaging and Modeling.* 2003.

25. ISO/TS, *Guide to the estimation of uncertainty in Geometrical Product Specifications (GPS) measurement, in calibration of measurement equipment and in product verification.* 1999.

26. ISO/TC, *Uncertainty of measurement - Part 3: Guide to the expression of uncertainty in measurement* in *Guide 98-3.* 2008.

27. Taylor, B.N. and C.E. Kuyatt, *Guidelines for Evaluating and Expressing the Uncertainty of NIST Measurement Results* 1994, NIST Technical Note 1297,.

28. Marino, R.M., et al. *High-resolution 3D imaging laser radar flight test experiments.* in *SPIE Conference on Laser Radar Technology and Applications* 2005. Orlando, FL: SPIE.

29. Donnelly, J.P., et al. *1-μm Geiger-Mode Detector Development.* in *SPIE Conference on Laser Radar Technology and Applications.* 2005. Orlando, FL.

30. Yuan, P., et al. *32 x 32 Geiger-mode LADAR cameras.* in *XV Laser Radar Technology and Applications.* 2010. Orlando, FL: SPIE.

31. Cheeseman, P., et al., *Super-Resolved Surface Reconstruction From Multiple Images.* 1994, NASA Technical report FIA-94-12, .

32. Stoker, C.R., E. Zbinden, and T.T. Blackmon, *Analyzing Pathfinder data using virtual reality and superresolved imaging.* Journal of geophysical Research, 1999. **104**(E54): p. 8889-8906.

33. Stettner, R., H. Bailey, and R. Richmond. *Eye-safe laser radar 3D imaging.* in *Laser Radar Technology and Applications IX* 2004. Orlando, FL: SPIE.

34. Stettner, R., H. Bailey, and S. Silverman. *Large format time-of-flight focal plane detector development.* in *Laser Radar Technology and Applications X* 2005. Orlando, FL: SPIE.

35. Buxbaum, B. and T. Gollewski, *Feasibility Study for the Next-Generation Laser Radar Range Imaging Based on CMOS PMD Technology.* 2002, Special Contract Report to NIST, Gaithersburg, MD.

36. Smithpeter, C.L., et al. *Miniature high-resolution laser radar operating at video rates.* in *Laser Radar Technology and Applications V.* 2000. Orlando, FL, USA: SPIE.

37. Schwarte, R., et al., *A new active 3D-Vision system based on RF-modulation interferometry of incoherent light.* SPIE Proceedings, 1995. **2588**.

38. Stann, B.L., W.C. Ruff, and Z.G. Sztankay, *Intensity-modulated diode laser radar using frequency-modulation/continuous-wave ranging techniques.* Optical Engineering, 1996. **35**(11): p. 3270-3278.

39. M.I.Skolnik, *Introduction to Radar Systems.* 2001: McGraw-Hill (The de-facto radar introduction bible.).

40. Stann, B.L., et al. *MEMS-scanned ladar sensor for small ground vehicles.* in *SPIE Conference on Laser Radar Technology and Applications* 2010. Orland, FL: SPIE.

41.   Kammerman, G.W., *Laser Radar*, in *The Infrared and Electro-Optical Systems Handbook*, C.S. Fox, Editor. 1996, SPIE Optical Engineering Press: Bellingham, WA.

42.   Gatt, P., *Range and Velocity Resolution and Precision and Time-Bandwidth Product*. 2003, Coherent Technologies Inc White Paper: Lafayette, CO.

43.   Karamata, B., et al., *Spatially incoherent illumination as a mechanism for cross-talk suppression in wide-field optical coherence tomography*. Optics Letters 2004. **29**(7): p. 736-738.

44.   Karamata, B., et al., *Multiple scattering in optical coherence tomography. I. Investigation and modeling*. J. Opt. Soc. Am. A, 2005. **22**(7): p. 1369-1379.

45.   Karamata, B., et al., *Multiple scattering in optical coherence tomography. II. Experimental and theoretical investigation of cross talk in wide-field optical coherence tomography*. J. Opt. Soc. Am. A, 2005. **22**(7): p. 1380-1388.

46.   Drexler, W., et al., *In-vivo ultrahigh-resolution optical coherence tomography*. Optics Letters, 1999. **24**(17): p. 1221-1223.

47.   Fercher, A.F., et al., *A thermal light source technique for optical coherence tomography*. Optics Communications, 2000. **185**: p. 57-64.

48.   Leitgeb, R., C.K. Hitzenberger, and A.F. Fercher, *Performance of fourier domain vs. time domain optical coherence tomography*. Optics Express, 2003. **11**: p. 889-894.

49.   Choma, M.A., et al., *Sensitivity advantage of sweep source and Fourier domain optical coherence tomography*. Optics Express, 2003. **11**(18): p. 2183-2189.

50.   Coddington, I., et al., *Rapid and precise absolute distance measurements at long range*. Nature Photonics, 2009. **3**: p. 351-356.

51.   Reibel, R.R., et al., *Ultrabroadband optical chirp linearization for precision length metrology applications*. Optics Letters, 2009. **34**: p. 3692-3694.

52.   Smartiehs. *http://www.ict-smartiehs.eu:8080/SMARTIEHS/publications*. 2010; Available from: http://www.ict-smartiehs.eu:8080/SMARTIEHS/publications.

53.   Beer, S., et al. *Smart Pixels for Real-time Optical Coherence Tomography*. in *Proc. of SPIE-IS&T Electronic Imaging*. 2004.

54.   Buttgen, B., *Extending Time-of-Flight Optical 3D-Imaging to Extreme Operating Conditions*, in *Faculty of Science of the University of Neuchatel*. 2006.

55.   Juberts, M., A. Barbara, and S. Szabo, eds. *Advanced LADAR for Unmanned Ground Vehicles*. Intelligent Vehicle Systems: A 4D/RCS Approach, ed. R. Madhavan, E.R. Messina, and J.S. Albus. 2007, Nova Publishers.

56.   Marino, R.F., et al. *A compact 3D imaging laser radar system using Geiger-mode APD arrays: system and measurements*. in *Laser Radar Technology and Applications VIII* 2003. Orlando, FL: SPIE.

57.   Cannata, W.R., et al. *Obscuration measurements of tree canopy structure using a 3D imaging ladar systems*. in *Defense and Security*. 2004. Orlando, FL: SPIE.

58.   Aull, B.F., et al., *Geiger-Mode Avalanche Photodiodes for Three-Dimensional Imaging*. Lincoln Laboratory Journal, 2002. **13**(2): p. 335-350.

59.   Scholles, M., et al. *Design of miniaturized optoelectronic systems using resonant microscanning mirrors for projection of full-color images*. in *Current Developments in Lens Design and Optical Engineering VII* 2006. San Diego, CA: SPIE.

60.   Siepmann, J.P. *Fusion of current technologies with real-time 3D MEMS ladar for novel security and defense applications*. in *Laser Radar Technology and Applications XI*. 2006.

61. Siepmann, J.P. and A. Rybaltowski, *Integrable Ultra-Compact, High-Resolution, Real-Time MEMS ladar for the Individual Soldier*, in *Military Communications Conference MILCOM*. 2005. p. 1-7.

62. Stann, B.L., et al. *Low-cost, compact ladar sensor for small ground robots*. in *Defense & Security*. 2009. Orlando, FL: SPIE.

63. Beck, J., et al. *Gated IR imaging with 128 × 128 HgCdTe electron avalanche photodiode FPA*. in *Proceedings of Infrared Technology and Applications XXXIII*. 2007: SPIE.

64. Beck, J., et al. *The HgCdTe Electron Avalanche Photodiode*. in *Infrared Detector Materials and Devices* 2004. Denver, CO: SPIE.

65. Kinch, M.A., et al., *HgCdTe Electron Avalanche Photodiodes*. Journal of Electronic Materials, 2004. **33**(6): p. 630-639.

66. Beck, J.D., et al. *MWIR HgCdTe avalanche photodiodes*. in *Materials for Infrared Detectors*. 2001.

67. McIntyre, R.J., *Multiplication Noise in Uniform Avalanche Diodes*. IEEE Transactions on Electron Devices, 1996. **13**: p. 164-168.

68. Beck, J., et al., *The HgCdTe Electron Avalanche Photodiode*, in *IEEE LEOS Newsletter*. 2006. p. 8 - 12.

69. CTC & Associates LLC, *LIDAR Applications for Transportation Agencies*, in *Transportation Synthesis Report*. 2010, Wisconsin DOT, .

70. Jennings, G. *Down in the dirt*. 2008; Available from: http://www.janes.com/.

71. Zhu, X., P. Church, and M. Labrie, *Chapter 13: AVS LIDAR for Detecting Obstacles Inside Aerosol*, in *Vision and Displays for Military and Security Applications*. 2010, Springer Science+Bussines Media.

72. Zhu, X., P. Church, and M. Labrie. *Lidar for obstacle detection during helicopter landing*. in *Laser Radar Technology and Applications XIII*. 2008: SPIE.

73. Harris, S. *Detecting threats to avoid trouble*. in *Europe Security & Defense SPIE Symposium*. 2008. Cardiff, UK.

74. Wall, R., *Looking Ahead: EADS sets its sights on an enhanced vision system using its helicopter laser radar system*, in *Aviation Week and Space Technology*. 2007.

75. Venot, Y.C. and P. Kielhorn, *A Dual-Mode Sensor Solution for Safe Helicopter Landing and Flight Assistance*. Microwave Journal, 2008. **51**(1).

76. National Institute of Building Science. *National BIM Standard, http://www.buildingsmartalliance.org/index.php/nbims/about/*. 2010 10/26/2010]; Available from: http://www.buildingsmartalliance.org/index.php/nbims/about/.

77. Akeel, H.A., *Robotic Technology for the 21st Century*, in *30th ISR Conference*. 1999: Tokyo, Japan.

78. Bostelman, R., et al., *Industrial Autonomous Vehicle Project Report*. 2001, NIST IR 6751

79. Dario, P. and R. Dillman, *EURON Robotics Technology Roadmap*. 2004.

80. EUROP_Executive_Board_Committee, *Strategic Research Agenda*. May 2006.

81. Bekey, G., et al., *International Assesment of Research and Development in Robotics*. 2006, World Technology Evaluation Center, Inc.

82. Energetics, *Robotics in Manufacturing Technology Roadmap*. 2006.

83. Bishop_Consulting, *Generation-After-Next AGV Technology Program: Final Report*. 2006.

84. Hong, T.H., et al., *Dynamic Perception Workshop Report: Requirements and Standards for Advanced Manufacturing*. 2010, NIST IR 7664, .

85.   NIST, *An assessment of the United States Measurement System: Addressing Measurement Barriers to Accelerate Innovation.* 2007, NIST Special Publication 1048, .

86.   Vandapel, N. and M. Hebert, *Finding Organized Structures in 3D LADAR Data*, in *IEEE/RSJ International Conference on Intelligent Robots and Systems.* 2004. p. 786 - 791.

87.   Hebert, M. and N. Vandapel. *Terrain Classification Techniques From Ladar Data For Autonomous Navigation.* in *Collaborative Technology Alliance Conference.* 2003.

88.   Bodt, B., et al. *Performance Measurements for Evaluating Static and Dynamic Multiple Human Detection and Tracking Systems in Unstructured Environments.* in *PerMIS.* 2009. Gaithersburg, MD

89.   Bodt, B. and R. Camden. *Detecting and Tracking Moving Humans from a Moving Vehicle.* in *SPIE 10th Conference on Unmanned Systems Technology.* 2008. Orlando, FL: SPIE.

90.   Szabo, S., J. Falco, and R. Norcross, *An Independent Measurement System for Testing Automotive Crash Warning Systems.* 2009, NIST IR 7545, .

91.   Toth, C.K., *R & D of Mobile LIDAR Mapping and Future Trends*, in *ASPRS Annual Conference.* 2009: Baltimore, MD.

92.   Jacobs, G., *3D Scanning: Laser Scanner Versatility Factors, Part 1.* Professional Surveyor Magazine, 2009.

93.   Cary, T., *Lidar Market: Status and Growth Trends*, in *International Lidar Mapping Forum.* 2009.

94.   Frost and Sullivan, *European Large Scale Micro Metrology Market.* March 2009.

www.ingramcontent.com/pod-product-compliance
Lightning Source LLC
Chambersburg PA
CBHW080248180526
45167CB00006B/2457